Where Have All the Soldiers Gone?

Where Have All the Soldiers Gone?

THE TRANSFORMATION *of* MODERN EUROPE

James J. Sheehan

MARINER BOOKS
HOUGHTON MIFFLIN HARCOURT
Boston • New York

First Mariner Books edition 2009

Copyright © 2008 by James Sheehan

For information about permission to reproduce selections from this book, write to Permissions, Houghton Mifflin Harcourt Publishing Company, 215 Park Avenue South, New York, New York 10003.

www.hmhbooks.com

Library of Congress Cataloging-in-Publication Data
Sheehan, James J.
  Where have all the soldiers gone? : the transformation
of modern Europe / James J. Sheehan.
    p. cm.
  Includes bibliographical references and index.
  ISBN-13: 978-0-618-35396-5
  ISBN-10: 0-618-35396-8
  1. Europe—History—20th century. 2. Europe—Economic
conditions—1945– 3. Europe—Defenses. 4. War and society
—Europe. 5. Politics and war. I. Title.
  D425.S54 2008
  940.55—dc22   2007009418

ISBN 978-0-547-08633-0 (pbk.)

Book design by Melissa Lotfy

Printed in the United States of America

DOC 10 9 8 7 6 5 4 3 2

Lines from "September 1, 1939," copyright 1940 and renewed 1968 by W. H. Auden, from *Collected Poems* by W. H. Auden. Used by permission of Random House, Inc.

For Natasha, Alexandra, and Sally

Contents

Politics will, to the end of history, be an area where conscience and power meet, where the ethical and coercive factors of human life will interpenetrate and work out their tentative and uneasy compromises.

—REINHOLD NIEBUHR,
Moral Man and Immoral Society

Acknowledgments

I began thinking about this book in the spring of 2001, when I had the good fortune to be a fellow at the American Academy in Berlin. Most of it was written in 2004 and 2005 while I was a fellow at the Stanford Humanities Center; the final revisions were done at the Wissenschaftszentrum in Berlin during the summer of 2006. I am grateful to Gary Smith at the American Academy, John Bender at the Humanities Center, and Jürgen Kocka at the WZB, as well as to the staffs of these three splendid institutions for their hospitality and generous assistance. I am also indebted to the John Simon Guggenheim Foundation, the Paul Davies family, and the School of Humanities and Sciences at Stanford for their financial support.

Daniela Blei, Jesse Kauffman, and Megan Wilcox-Fogel provided valuable research assistance. My friends and colleagues David Kennedy and Norman Naimark read an earlier version of the book and offered many helpful suggestions. Don Lamm, my agent, was involved with this project from the beginning to the end; without his help this book would not exist. At Houghton Mifflin, Amanda Cook was a splendid editor — tough and relentless, encouraging and supportive. Her influence can be found on every page. Will Vincent did yeoman work during that long twilight journey from manuscript to book.

Those who know us will not be surprised to learn that my greatest debt is to Margaret Lavinia Anderson, for being my ideal reader, resident critic, scholarly model, and so much else.

Berkeley, California
December 2006

Prologue

War and Peace in the Twentieth Century

O N SATURDAY, February 15, 2003, the largest demonstration in European history was held to protest the impending war against Iraq. In London, an estimated million people overflowed Trafalgar Square, filling the city's streets from the Thames embankment to Euston Station; a million marched in Barcelona and in Rome, 600,000 in Madrid. A half million braved the freezing cold in Berlin's Tiergarten, almost as many as usually attended the Love Parade held there in the summertime. Everywhere the crowds were peaceful. There were few arrests, no violence. The demonstrations attracted a rich variety of participants: there were some tough-looking adolescents in leather and young people wearing Palestinian head scarves or anarchist black, but the overwhelming majority were respectable citizens in warm winter coats and sensible shoes—pensioners, middle-aged academics, union members, high school and college students. There were lots of families, parents and grandparents who had not marched since the sixties, children experiencing for the first time a political demonstration's distinctive blend of exhilaration and discomfort. One German newspaper called it "an uprising of ordinary people."

Many of the demonstrators carried banners and placards, some prepared by the organizers, others homemade, which proclaimed the various motives that had brought them into the streets: "Freedom for Palestine," "No Blood for Oil," "Stop Mad Cowboy Disease," "America, the Real Rogue State," "Make Tea, Not War," and

Saying no to war. Demonstrators at Nelson's Column,
Trafalgar Square, February 15, 2003.

(my personal favorite) "Down with This Sort of Thing." Unlike the demonstrations against the war in Vietnam, no one showed any sympathy for the other side; there were no Iraqi flags or pictures of Saddam Hussein. For most people, the real issue was not who was right or wrong, but whether war was the answer. Seventy-three-year-old Thomas Elliot, a retired solicitor from Basildon, Essex, explained why he was attending his first political demonstration: "I remember the war," he told a reporter, "and the effect the bombing had on London. War should only be used when absolutely necessary." In

Berlin, Judith Rohde and Ricarda Lindner, fourteen-year-old classmates from a local high school, were surprised that anyone needed to ask why they were marching. "War," they said, "is not a solution." Hilde Witaschek, at seventy-seven a veteran peace marcher, added, "We experienced war when Berlin was liberated—no more war, *nie wieder Krieg*." In city after city, when one looked out across the ocean of people, the sign that appeared most often contained a single word: "No."

Some observers regarded February 15 as a turning point in European history. Dominique Strauss-Kahn, a former French cabinet minister, declared that a new "European nation" had been born that day. A few months later, in an article originally entitled "February 15: What Unites Europeans," Jürgen Habermas and Jacques Derrida, two of Europe's best-known intellectuals, called on Europeans to "counterbalance the hegemonic unilateralism of the United States in the international arena and within the United Nations." Like Strauss-Kahn, Habermas and Derrida argued that Europe's opposition to American militarism could create a new European identity, an identity based, above all else, on a rejection of war as an instrument of national policy.

On February 5, just ten days before the great demonstrations, a book by Robert Kagan, *Of Paradise and Power: America and Europe in the New World Order*, was published. Kagan, who served briefly in the Reagan administration and was an early advocate of the use of American power to spread democracy in the world, had been among the first to push for war against Iraq. His book, which quickly found a place on the bestseller list, was based on an essay called "Power and Weakness," which had appeared the previous spring in a rather obscure journal, *Policy Review*. Kagan tried to summarize the differences between Europe and America by borrowing the title of a recent book about gender difference: "on major strategic and international questions," he declared, "Americans are from Mars and Europeans are from Venus." Transatlantic discord is not merely the result of Europeans' opposition to a single event or the policies of one particular administration. "It is time to stop pretending," Kagan wrote, "that Europeans and Americans share a common view of the world, or even that they occupy the same world." Europeans have turned away from power, preferring to live in a posthistorical paradise; Americans recognize that in the real world power and military might are still essential. "The reasons for the transatlantic divide are deep, long in development, and likely to endure."

Kagan's analysis, like Habermas and Derrida's call for a new European identity, reflected the passionate debates ignited on both sides of the Atlantic by the Iraq War. We shall return to these debates in this book's final chapter. But for the moment, it is sufficient to acknowledge the truth at the core of the comparisons all three made between Europeans and Americans: at the beginning of the twenty-first century, many more Americans than Europeans were prepared to accept the necessity of using violence to resolve international disputes. In 2003, when a poll by the German Marshall Fund asked Americans whether they believed that, under certain circumstances, war was necessary to obtain justice, 55 percent strongly agreed. In France and Germany, only 12 percent held that opinion.

Europe at the beginning of the twenty-first century is economically strong but uninterested in transforming this strength into military power. The power that European states do project internationally is economic, cultural, and legal, the outward expression of the values and institutions that matter most in their relations with one another and with their own citizens. By contrast, the United States operates on a global stage, with an enormous network of military bases, a thick web of alliances and agreements, a truly global sphere of influence and power. America has become what Timothy Garton Ash called "the last truly sovereign European nation-state." The ability and willingness to make war has traditionally been the essence of sovereign statehood. How this has changed, at least in Europe, is the subject of this book.

Well before the conflict in Iraq revealed the fissure in European and American relations, some scholars pointed to a declining belief in the efficacy of war, not simply in Europe but worldwide. In a book entitled *Retreat from Doomsday,* the American political scientist John Mueller maintained that major war—as opposed to civil strife and organized criminality—was becoming obsolete. According to Mueller, the values and assumptions that had once made war an inevitable part of human affairs were now dissolving; people no longer believed that war was an effective instrument of policy, that "victory" would ever be worth the price. War, therefore, was not an intrinsic part of human experience but would, like other apparently incurable social evils such as dueling and slavery, eventually fade away. When Mueller first presented his ideas in 1989, several commentators, especially students of war like Michael Howard and John Keegan, expressed skepticism that the subject, to which they had devoted a lifetime of distinguished scholarship, was on its way to historical oblivion. But in the course of the nineties, Keegan and Howard, together with many other well-informed observers, began to wonder whether the age-old connection between war and states might indeed be coming to an end.

This book will make two central arguments: first, the obsolescence of war is not a global phenomenon but a European one, the product of Europe's distinctive history in the twentieth century; second, the disappearance of war after 1945 created both a dramatically new international system within Europe and a new kind of European state.

We will see how the historical developments that made modern European wars so extraordinarily destructive were the very ones that ultimately banished, for the first time in Europe's long and bloody history, international violence from the European society of states. The democratization of politics and society, for example, gave European governments the capacity to mobilize human resources and to raise armies of unprecedented size and thus dramatically increased both the scale and the intensity of combat. But democratization also encouraged the conviction that ordinary people, those who—as always—bore the burdens of war, should have a say in when or if states should fight and that, given the choice between war and peace, they would chose the latter. Similarly, the growth of industrial production made it possible to make and deploy weapons of unparalleled destructive power. But industrialization also expanded the connections among peoples and nations, weaving a web of interdependent relationships that required and sustained peaceful exchange. A major war, many people realized, would damage, perhaps destroy, these relationships and thus inflict incalculable harm on European economic and social life.

In 1900, the European society of states was governed by men who recognized the potential risks of a European war. In order to manage these risks, statesmen maintained an elaborate set of institutions designed to preserve the peace or, should that fail, to contain international violence. We need not overestimate the effectiveness or the benevolence of this "concert of Europe." It left plenty of room for violence outside Europe and threats of violence within; it was always driven by self-interest and, like every international system, usually worked to the benefit of the strong and at the expense of the weak. Nevertheless, the international order that existed at the beginning of the twentieth century seemed to work remarkably well.

Significantly fewer Europeans died in combat during the nineteenth century than in the eighteenth, not to mention in the monumentally bloody twentieth. Between 1648 and 1789, the European powers had fought forty-eight wars, some of them, like the Seven Years' War in the mid-eighteenth century, lasting several years and stretching around the world. Between 1815 and 1914, there were only five wars in Europe involving two great powers; all of them were limited in time and space, and only one of them involved more than two major states. From the end of the Franco-Prussian War in 1871 until the outbreak of the Great War in 1914, the European states were at peace with one another. This was the longest period without war in European history until it was surpassed toward the end of the twentieth century.

During the long peace of the late nineteenth and early twentieth century we can find the historical roots of the civilian policies and institutions that would eventually dominate European public life. These policies and institutions were directed inward, toward domestic goals; they sought to encourage economic growth, promote commerce, and provide new kinds of services for their citizens. As in the period after 1945, these developments were inseparable from unprecedented economic expansion. Throughout the second half of the nineteenth century, European manufacturing and agricultural production increased dramatically. Despite a growing population, per capita income rose, as did gross domestic product. Growth was geographically uneven and its benefits unequally distributed, but by 1900 European society was becoming increasingly orderly, peaceful, and prosperous.

Although they lived in peace, Europeans at the beginning of the twentieth century constantly confronted the possibility of war. "The great powers of our time," the German chancellor Otto von Bismarck told a Russian diplomat in 1879, "are like travelers, unknown to one another, whom chance has brought together in a carriage. They watch each other, and when one of them puts his hand into his pocket, his neighbor gets ready his own revolver in order to be able to fire the first shot." No responsible statesman was prepared to let down his guard by looking away from his companions; the chance that one of them, accidentally or intentionally, might draw his weapon could never be dismissed. Preparing for war was the statesman's most important duty—not his only duty, to be sure, but the one that took precedence over all others. Economic prosperity, commercial vitality, and social welfare were worthwhile goals; all of them contributed to the state's power and stability; but they counted for nothing if the existence of the state was not secure. Security meant creating and sustaining the kind of army necessary to fight and win a modern war. As one German politician put it toward the end of the nineteenth century: "What good are the best social reforms if the Cossacks come?"

In the summer of 1914, the leaders of the great powers decided that they had no choice but to fight. Some of them may have actively sought a European war, but no one wanted the war they got, a war in which Europeans employed their extraordinary ability to mobilize human and material resources to destroy one another. This was a democratic war that reached into the lives of virtually every European; it was also an industrial war, in which death and devastation became the principal purpose of economic production. The war consumed millions of lives, most of them young, and vast resources, all of them wasted. It uprooted ancient institutions, disrupted newly created economic bonds, and shattered the delicate arrangements that had helped to restrain the great powers since 1815.

Surveying the wreckage left in the war's wake, many were convinced that major wars had indeed become obsolete; surely Europe could not survive another. But others drew a different lesson from the war. For them, peace had come too soon, before victory had been obtained, the enemy destroyed, society purged of its toxins. To these people, war was the source of the heroism, discipline, and comradeship from which a new political order could be built. Europe in the 1920s and 1930s was divided along many fault lines—between democracy and dictatorship, communism and capitalism, right and left—but the most important was between those who rejected and those who embraced political violence at home and abroad. In the end, the proponents of violence carried the day, plunging Europe into a second, yet more terrible war, in which, once again, the forces of democracy and industry were forged into weapons of mass destruction.

If a broad popular consensus about the futility of war were enough to guarantee peace, then one world war should have been sufficient. But unlike slavery or dueling, which could gradually fade away as its cultural support eroded, war would remain a danger as long as one state stood ready and willing to fight. *All* the strangers in Bismarck's metaphorical carriage had to be sure that none of their fellow travelers would reach for a weapon. Security was indivisible.

This indivisible sense of security arose in Europe after 1945, when the United States and the Soviet Union imposed a new order on the continent, dividing and organizing the European states in what became a remarkably stable and peaceful system. This system provided the incubator within which the states of western Europe were gradually transformed. They became civilian states, states that retained the capacity to make war with one another but lost all interest in doing so. The result was an eclipse of violence in both meanings of the word: violence declined in importance and it was concealed from view by something else—that is, by the state's need to encourage economic growth, provide social welfare, and guarantee personal security for its citizens. The eclipse of violence happened gradually. It was a slow, silent revolution, hidden in plain sight, but it was nonetheless a revolution as dramatic as any other in European history. In order to understand the character and significance of this revolution, we must turn to a time when war was still the most important element in the life of European states.

PART I

Living in Peace, Preparing for War, 1900–1914

I

* * *

"Without War,
There Would Be No State"

T HE CHAPTER TITLE comes from Heinrich von Treitschke's
lectures on politics, delivered at the University of Berlin in
the 1880s and 1890s. Like the students who crowded the university's
largest auditorium to hear him, Treitschke had in mind the example
of his own state, the united Germany that had been forged in three
recent wars. But he believed that all states depended on war for their
origins and existence. "Every state known to us," he insisted, "was
created by war; the protection of its citizens with weapons remains
the state's first and most essential task."

To confirm the truth of Treitschke's view that war and states were
inseparable, his listeners had only to step outside his lecture hall onto
Unter den Linden, the grand thoroughfare that ran from the Brandenburg Gate past the university to the royal palace. In front of the
university, opposite the history department library, was C. D. Rauch's
equestrian statue of Frederick the Great, the warrior king whose victories had ensured Prussia's status as a great power. Rauch portrays
Frederick in the simple officer's tunic that he habitually wore — he
was among the first European monarchs to adopt military dress as
his normal public attire, thereby emphasizing the army's practical
and symbolic importance as a source of royal authority. Like his fellow monarchs, Frederick played many roles — lawgiver, patron of the
arts, chief of the civil administration, and head of the church — but
he fulfilled these functions dressed as a soldier. His political, legal,

and cultural authority was dependent on, and inseparable from, his command over the army. Rauch left no one in doubt about the primacy of the king's military role: the base of his statue is dominated by the generals who had shared the perils and glory of Frederick's campaigns. Civilian figures were given a subordinate position; it did not pass unnoticed, for example, that the great philosopher Immanuel Kant occupied a place just beneath the tail of the monarch's horse.

Along Unter den Linden passed the military parades, like the one portrayed in Franz Krüger's painting of 1827, *Parade on the Opernplatz*, which records a visit of the Russian crown prince to Berlin. With the elite grenadier guards lining the street, the royal visitor, dressed in the uniform of the 6th Brandenburg Cuirassiers, rides at the head of a squadron of troops. In the corner of the painting a crowd of civilians, including the artist himself and several of his friends, watch with a respectful enthusiasm that unites a variety of social groups. Parades like this one were the usual way for states to welcome important guests, commemorate royal birthdays and weddings, and celebrate national holidays. Such ceremonies, the emperor William II pointed out around the turn of the century, did not simply evoke past glories but were themselves "tests of manly discipline, demonstrations of the individual's willingness to master his nerves and muscles, his ability to subordinate his own will to the collective." In other words, soldiers on parade vividly represented what the state demanded and expected of its subjects.

We often think of Berlin as having a peculiarly militarized civic culture, but military displays were also common in Vienna, where the movement of smartly dressed units through the streets was part of the city's everyday life. In the Habsburgs' capital, even so religious a holiday as Corpus Christi, the feast celebrating the Holy Eucharist, was marked by festivities in which ecclesiastical, dynastic, and military institutions supported and reinforced one another. Photographs from the turn of the century show the male members of the Habsburg family, all in military dress, marching behind the sacrament along streets lined with soldiers.

Military institutions also represented the state in republican France, despite the government's often troubled relationship with its army. In 1871, soon after a defeated France had signed the Treaty of Frankfurt with the Germans, the elderly Marshal MacMahon led 120,000 men as they marched across the plain of Longchamp in Paris. Evoking cheers from a huge crowd of spectators, the marshal embraced the civilian representatives of the new regime, thereby committing the military to support the republic. After 1880, when the republic adopted July 14, Bastille Day, as its national holiday, it celebrated the occasion with impressive military parades on Longchamp and eventually in every French town that could boast a garrison. When reviewing the troops in Chartres in 1894, President Casimir-Perier saluted what he called "this grand school of patriotism,

A warrior king. Monument to Frederick the Great, Berlin.

the army." Despite differences in the civic function of military institutions among the various European states, in all of them the army embodied what Casimir-Perier called "the image of the nation"—or, more accurately, the image of what the nation wanted to be.

Reminders of the state's military history were woven into the fabric of every major European city. When Napoleon III rebuilt Paris in the 1850s and 1860s, he named several major streets after battles —his uncle's brilliant victory over the Prussians at Jena in 1806, and his own army's rather lackluster performances at Sebastopol in 1855 and Magenta in 1859. Every capital had its victory monuments: Nelson's Column in London's Trafalgar Square, which celebrates Britain's great naval victory over France in 1805; the Arc de Triomphe in Paris, begun by the first Napoleon at the height of his power in 1806 as a memorial to his Grand Army; the Siegessäule in Berlin, built between 1869 and 1873 to mark Prussia's successes in the wars of German unification.

The tombs of national heroes adorned every capital: from Wellington's modest granite casket in St. Paul's to Napoleon's elaborate resting place in the crypt of the gold-domed Invalides. In addition to these national shrines, thousands of more modest monuments marked the nation's martial history: the simple cenotaphs in Prussian towns that honored the local men who had fallen at Königgrätz or Sedan, and the plaques on the walls of small churches in isolated Scottish villages, inscribed with the names of men who died in some distant outpost of empire. The military was grand and distant, connected to great deeds and powerful men, but also familiar and close, part of everyday life.

Every modern state is an imagined community, since a state is too large and complex to be experienced directly. That is why, as Michael Walzer reminds us, a nation "must be personified before it can be seen, symbolized before it can be loved, imagined before it can be conceived." European states have always had to shape their citizens' political imagination with whatever cultural symbols and historical memories were available. At the beginning of the twentieth century, these symbols and memories had a distinctly military cast. States wanted, above all, to be identified with the heroism, self-sacrifice, and duty that had made their victories possible and their defeats endurable. Men in uniform personified the virtues on which the state's existence depended, just as the army and navy symbolized its discipline and cohesion. Without the capacity to make war, the early-twentieth-century state could not exist—indeed, it could not even be imagined. This is why every country, no matter how small or vulnerable, had an army of its own.

War was deeply inscribed on the genetic code of the European state: "States make war," as the American sociologist Charles Tilly concisely put it, "and vice versa." But while war and states have always been entwined, the nature of their relationship constantly changes. By 1900, two developments had transformed both war and states. First, both were democratized: the emergence of mass reserve armies engaged a far greater portion of the population than most people had previously thought necessary or desirable. Second, both were profoundly affected by industrialization: the application of technology to warfare made it possible to have larger, more complex, and more expensive armies than ever before. States had always made war, and war had always made states—but the kind of war that states were preparing to fight at the beginning of the twentieth century was without historical precedent.

The democratization of war was one of the results of the great political revolutions at the end of the eighteenth century. But despite the mass revolutionary army's political power and military effectiveness, most European states had been reluctant to arm and train their populations. Once Napoleon had been defeated, therefore, most states returned to the relatively small, largely professional armies on which they had traditionally depended. They raised these armies with a socially discriminatory mode of conscription, which placed the burden of national defense on the sons of the poor, who spent anywhere from six to twenty years in uniform.

Except for the conscripts themselves, virtually everyone liked this system. The well-to-do were delighted that they and their sons could avoid military service. In France and Italy, for instance, anyone who got a "bad number" in the conscription lottery had the right to hire a substitute to take his place; in France, up to a quarter of each year's recruits were substitutes, frequently veterans who accepted payment as a sort of bonus for reenlisting. The leaders of the army were pleased to have men who had been hardened, drilled, and disciplined by long years of service. And governments were reassured that, because their soldiers lacked strong ties to civilian life, they would be ready and willing to defend the established order against domestic unrest as well as a foreign invader. The fate of a regime, as the era of revolution had demonstrated, might depend on the reliable conduct of troops when confronted by an incendiary mob.

European states abandoned these habits only because they became convinced that their survival demanded large-scale military reforms. The somewhat unlikely source of this pressure for change was Prussia, which, since its consolidation as a state in the late seventeenth century, had generally been regarded as the weakest of the great powers. Prussia almost disappeared from the map when it was defeated by Napoleon in 1806; it recovered enough to play a respectable if subordinate role in the emperor's final defeat. After 1815, Prussia had been the only major power to retain universal military service: two years of active duty, five years in the reserves, eleven more in the militia. But budgetary constraints kept the number of recruits relatively small, filled the militia with poorly trained troops, and put the army's military effectiveness and political reliability in doubt. Moreover, the Prussian army lacked combat experience. It had not participated in either the Crimean or the Italian wars, nor had it been tempered by the colonial combat in which British, French, and Russian soldiers were continually engaged. A French observer of the Prussian army's maneuvers in 1861 was decidedly unimpressed by what he saw. "They are," he reported, "an embarrassment to the profession."

By 1861, however, Prussian military institutions were being radically transformed. Three years earlier, Prince William, who had spent most of his adult life in the regular army, became regent when his elder brother was incapacitated by a series of strokes. Aided and encouraged by Albrecht von Roon, the minister of war, William set out to reform the army by lengthening the term of active service to three years, tightening control over the reserves, reducing the role of the militia, and introducing a number of changes in organization, training, and equipment. Helmuth von Moltke, who had become the chief of the General Staff in 1857, was one of the few officers in Europe to recognize the military application of railroads and to understand how to use them to deploy troops swiftly and efficiently.

More than any other single innovation, Moltke's strategic exploitation of the railroad represented the fusion of technology and war that would do so much to shape the modern world.

Prussia's skillful and ambitious minister president, Otto von Bismarck, was willing to use military means to dominate the smaller German states because he believed that this reformed army could achieve quick and decisive victories on the battlefield. Between 1864 and 1871, Prussia fought three brief and successful wars against Denmark, Austria, and France. That the Prussians (allied with Austria) easily defeated Denmark surprised no one, but their victory over Austria after a single battle and only seven weeks of hostilities startled the other powers, as did their somewhat more prolonged but no less triumphant war against France.

What made these victories possible? The Prussian army's strategy and tactics, most experts agreed, were not significantly superior to either Austria's or France's. Nor did Prussia have an obvious advantage in weapons. When properly led, the soldiers in every army had fought with courage and tenacity. In the end, victory was the result of better preparation, planning, and organization, which had allowed Prussia to mobilize more men, move them more quickly, and equip them more efficiently. In 1903, Ferdinand Foch, a future marshal of France, summarized the lessons to be learned from Prussia for his students at the French war college: "mass" and "preparation," he said, were the essential sources of victory in modern war.

In order to emulate the Prussians, therefore, two kinds of innovations were necessary. First, states had to have mass standing armies (Prussia and its German allies sent more than a million men into France in 1870), which would require conscripting a large percentage of the male population. Because so many men could not be kept out of the workforce for a prolonged period, they would have to return to civilian life after two or three years of active duty, but would still remain part of a ready reserve. Second, states had to be prepared to deploy these mass armies swiftly and effectively. The Prussians had transformed the tempo as well as the scale of warfare: modern communication, and especially the railroad, made fast deployment possible, but it also posed profound organizational difficulties. In 1870, France had not recovered from the confusion surrounding its initial mobilization; because of poor planning, the French army's system of communication had broken down, men and their equipment often ended up in different places, trains arrived at their destinations without plans to unload them. Before they could remedy these mistakes, the French forces had been soundly and decisively beaten.

While it was clear enough that states needed a well-prepared mass army, creating it required overcoming formidable objections—from the officer corps, who wanted fewer but better soldiers with longer enlistments; from the middle classes, who wanted to protect their exemption from active service; from left-wing politicians, who feared a militarization of society; and from many ordinary taxpayers, who worried about skyrocketing military budgets. What the great Prussian strategist Carl von Clausewitz said about combat turned out to be equally true of getting ready for war: it was simple to know what needed to be done; doing it, however, was extremely difficult.

In no European state was the imperative of military reform more pressing, or more difficult, than in France, which had been the chief victim of the harsh lessons taught by Prussia in 1870. In 1872, the new republic passed a revised conscription law that increased the number of those in each age cohort who were eligible to be drafted. Even so, a lottery still decided whether someone spent one year or five in the regular army. And while it was no longer possible to buy a substitute, a number of exemptions continued to favor the prosperous and educated; in practice, only young men who could not afford to stay in school were likely to serve five years. In 1889, over strong objections from the officer corps, the length of service was reduced to three years. The law regulating military service was fundamentally changed in 1905, when nearly all French males became liable for two years in the regular army. In order to make up for a declining population and to match the expansion of the German army, service was increased to three years in 1913.

Every European state faced a slightly different set of impediments to military reform. In France, resistance came from the reluctance of prosperous families to assume their responsibility to serve in the army and from the mutual distrust between the republican government and the conservative, heavily monarchical officer corps. Italy's conscription law had even more exemptions than France's, most notably a free ride for a family's only son. Potential recruits who met minimal educational requirements and could afford to buy their own uniforms had the right to serve only one year. Most of these exemptions were not abolished until 1909, when the term for active duty was set at two years for everyone.

Austria's problems were national and economic rather than social and ideological: although the Austrian army remained an imperial institution in which orders were given in German, the reserves were divided between German and Hungarian units. Financial constraints and the political opposition of the Hungarian parliament limited the expansion of the armed forces, which remained undersized and ill equipped. Despite the social prestige officers enjoyed, low pay made a military career decreasingly attractive for young men with other options.

Like the other powers, Russia's army reforms, introduced in the 1870s, reduced the length of service, in the Russian case to six years (before 1861, it had been twenty-five years, reduced in the 1860s to eight). Although attempts were made to introduce more humane disciplinary measures and better living conditions, most ordinary soldiers continued to be poorly fed, shoddily equipped, and harshly treated (Russia spent about half as much per soldier as did Germany). There was, therefore, a good deal of truth in Friedrich Engels's critical comment that the Russians "have adopted a system of universal liability for which they are not civilized enough."

Britain was the one great power that did not try to create a mass army based on short-term conscription and ready reserves. Like the other powers, Britain had introduced some reforms in the 1870s, which reduced the length of voluntary enlistment from twenty years to twelve (six on active duty, six in the reserves) and abolished the practice of buying and selling commissions. The insufficiency of these changes was revealed during the Boer War, in which Britain prevailed, but only at great cost. "When we speak of the critical military condition of England," a high-ranking Austrian officer told his colleagues in 1900, "we do not refer to that of the army engaged in South Africa, but to the fact that, by this employment, England was all but denuded of troops."

The size of the army remained the central issue when Richard Haldane, a lawyer, civic reformer, and educational innovator, became secretary of state for war in December 1905. "Our main object," he declared a month later, "must be the education and organization of the nation for the necessities of imperial defense." Nevertheless, Haldane refused to abandon the voluntary principle on which British defense policy had traditionally rested. He remained persuaded that only a professional army could be readily employed in the empire; conscripts would have neither the will nor the ability to fight distant battles in defense of imperial rule. The security of the British Isles themselves would remain the responsibility of the Royal Navy and what Haldane hoped would be a greatly expanded force of territorial reserves. After an encouraging start, however, enlistment in the territorial units began to fall off, which increased pressures for conscription. Low numbers convinced Charles à Court Repington, the *Times* military correspondent, that "the turn of compulsion will come."

By the beginning of the twentieth century, every European state, with the exception of Britain (and, of course, that perennial exception to all generalizations about Europe, Switzerland), had established a mass reserve army. This meant that an increasing number of young men had some sort of military training: in 1870, about 1 in 74 Frenchmen and 1 in 34 Germans were ready for action; by 1914, it was 1 in 10 and 1 in 13. Russia, with its vast population, trained 35 percent of its males of military age. France, which faced a demographic crisis, trained 85 percent, Germany 50 percent, Austria-Hungary 49 percent.

Per capita, the largest armies in Europe were in the Balkan states, where, as we will see, war remained a real possibility: tiny Montenegro, with a population of about 250,000 in 1909, could put into the field an army of 30,000 to 40,000, which included virtually every male between eighteen and sixty-two. Bulgaria, where most able-bodied men remained in the reserves until they were forty-six, had an army of 350,000. "We have become," a Bulgarian general proudly proclaimed in 1910, "the most militaristic state in the world." Fully mobilized, the Russian army had 3.4 million men, the Germans 2.1 million, the French 1.8 million, and the Austrians 1.3 million. Should a war among the European powers break out, armies of unparalleled size would be deployed against one another, with unprecedented speed. In the meantime, a significant part of the male population and a considerable amount of each state's revenues were directed toward preparing for this possibility.

• • •

The mass reserve army made military service a part of the life experience of millions of European men and gave military institutions a central place in European society. To recruit, train, equip, supply, and deploy millions of citizen soldiers required an array of administrative agencies, complex and expensive equipment, and an elaborate infrastructure. Only a handful of the world's states were sufficiently rich and well organized to raise and maintain this kind of army—which is why the list of great powers remained so stable between the end of the eighteenth century and the beginning of the twentieth. With the single and significant exception of Japan, no non-Western state was able to build a modern army until well into the twentieth century.

All the powers spent enormous sums of money on their armed forces in the two decades before 1914, and especially after 1912, when an intensified arms race reflected the deteriorating international situation. The European states could afford this kind of expenditure because they had prosperous and dynamic economies. As a percentage of their net national product, therefore, the growth in defense spending was sizable but much less dramatic than the absolute numbers: for Britain, for example, expenditures grew from 2.5 to 3.2 percent between 1893 and 1913, despite a massive campaign of naval construction. Nevertheless, in every state the military budget was a source of political controversy, both because of its size and because of the tax structure on which it was based. In the end, the armed forces usually got most of what they wanted, but not without serious debate and dissent. Even in Germany, with its precarious geopolitical position and traditionally powerful military institutions, parliamentary forces and civilian administrators set limits on the army's ambitious program for expansion.

In order to build and maintain mass reserve armies, states not only needed money, they also had to be able to measure, count, and if necessary coerce their populations. Obviously, states had to know who their potential recruits were and where they could be found—information that was sometimes more difficult to gather than it might seem. Some French villages, for instance, reported demographically improbable preponderances of female births; the names of male infants somehow did not find their way into the official records. Migration, either to another region or, in the case of Italy and Russia, overseas, became a favorite means of avoiding military service. An

Italian law of 1888 that prohibited men below the age of thirty-two from leaving the country was largely ineffective. Sometimes young men simply did not show up when they were called: in one French town, only one out of a cohort of eighteen recruits was present when the time came to begin military service. Nobody seemed to know where the others had gone.

Conscription required that states define and regulate who belonged and who did not. In Prussia, military considerations had been behind a law issued in 1842 that shifted the power to establish who was a Prussian subject—and therefore who was eligible for military service—from the local community to the state. Prussian subjects were free to move around the kingdom, but those of military age could not emigrate without permission. Conscription played a role in the French law of 1889 that conferred citizenship—attended, of course, by eligibility for military service—on the children of resident foreigners. Otherwise, the government realized, aliens would have an unfair and lasting advantage over natives. The same reasoning encouraged the French government to grant certain benefits to French citizens that were not available to foreign residents. The more universal military service became, therefore, the more tightly the strands of identity and obligation, welfare and duty, were knitted in a single category of citizenship.

States also needed ways to decide which citizens were eligible and fit to serve in their armies. This required uniform physical examinations: potential recruits' height and weight were measured, their eyesight tested, their medical condition examined for signs of chronic disease or mental instability. The authorities had to evaluate requests for exemptions for reasons of health or special hardship. It was also necessary to monitor the categories of service that were offered to particular groups. If, for example, a young man's educational status gave him certain privileges, then a government agency had to be sure his claim to this status was legitimate. In Prussia, where graduates of academic secondary schools could apply to limit their active service to one year and then to become officer candidates in the reserves, the administration certified which schools met the graduation requirements. Identity cards, medical examinations, educational standards, and a variety of other forms of government regulation were directly connected to the creation of a mass reserve army.

*Her er i hvert og
reserve*

The state's penetration of society was also deepened by the extended network of institutions necessary to register, train, and keep in touch with reservists. Those in the active reserves had to appear for regular training. In case of war, reservists had to be contacted, pulled from their everyday lives, brought together in a designated place, provided with the proper equipment, then transported to a theater of operations, where their units would be coordinated with hundreds, perhaps thousands of others. In order for a mobilization to run smoothly, elaborate plans, preparations, instructions, and supplies had to be in place. Coded orders had to be written, railroad timetables prearranged, depots filled with equipment, and kitchens, aid stations, and command posts made ready for immediate operation. The larger the army and the more technologically complex its equipment, the more difficult this organization became and the greater the potential for confusion, error, or delay.

At the core of every mass reserve army was a cadre of professionals: the officers, noncommissioned officers, and long-term enlisted men on whom the institution's effectiveness ultimately depended. Professionals provided strategic planning and direction—usually in a General Staff created on the Prussian model—and trained the steady influx of conscripts. Recruiting, educating, and motivating these professionals also absorbed the state's resources and organizational capacity. Giving former soldiers financial support, medical care, and other benefits was often the first, and in some places the only, manifestation of the state's welfare mission. Here again, the state's military obligations led to an expansion of its power to shape the rest of society. *Her haft Meget indflyox*

The central purpose of this complex military machine was to transform civilians into soldiers. This is a task all armies face, except in those societies where every male is reared to be a warrior. Until the end of the eighteenth century, European armies were made up of the unskilled, the unwilling, and the unlucky, who were turned into soldiers by endless drill, brutal discipline, and the threat of draconian punishment. Such methods were not suitable for the mass reserve army, whose ranks were filled with citizen soldiers. In the first place, military institutions were under constant public scrutiny; instances of the sort of institutionalized brutality that had pervaded traditional armies now produced outraged articles in the press and embarrassing questions in parliaments. Politicians constantly called for the abolition of corporal punishment and the reform of military justice. Most professional soldiers grumbled about meddling from uninformed civilians, but they recognized that there were limits to what could be done to men who came from, and would soon return to, civilian life. Soldiers, Marshal Lyautey acknowledged, had to be treated not as brutes but as Frenchmen.

Of course, no army functions without coercion. After all, conscripts had to be taught how to fight—that is, they had to acquire the knowledge, skills, and stamina needed to survive in combat. This meant abandoning the habits of mind and spirit that they had brought with them from civilian life. The new uniforms, the close-cropped hair, and the other rough rituals of barracks life were designed to make clear to the recruits that they had entered a new world where it was essential to obey their superiors without hesitation, support their comrades without qualification, pursue their enemies without mercy. "Discipline," noncommissioned officers in the French army were taught, "will be the soldier's religion . . . The regiment is the school of subordination, of the virile spirit, of male pride."

But while professional soldiers insisted that subordination was essential and discipline was the soldier's main virtue, most experts recognized that modern war could no longer be based on the kind of unquestioning obedience that had been necessary to compel traditional infantrymen to remain in rank and continue to fire even as their comrades fell around them. New, more lethal weapons demanded looser formations, greater tactical mobility, more individual initiative. The twentieth-century soldier's discipline could not be imposed with endless drill and brutal treatment; it had to be the product of indoctrination, inspiration, and affection.

Like the states they defended, modern armies required more than disciplined passivity. Citizen soldiers were supposed to fight not from fear but from devotion to their country and commitment to their comrades. Members of these new armies had to be convinced that the security, perhaps even the existence, of their states relied on their willingness to fulfill their military obligations, answer the call to arms, obey orders to kill, and if necessary to die.

Because the modern soldier was supposed to act from conviction rather than compulsion, it was essential that he feel part of the nation, that he be an active citizen rather than a passive subject. This is why conscription was often associated with social and political emancipation. The French reactionary theorist Hippolyte Taine called it the "twin brother" of universal suffrage, "both of them the blind and terrible guides or masters of the future." At the other end of the political spectrum, Friedrich Engels, the cofounder of modern socialism and an expert in military matters, wrote that "contrary to appearance, compulsory military service surpasses general franchise as an agent of democracy." Taine and Engels were right: from the moment the "nation in arms" emerged during the revolutionary wars of the 1790s, universal military service was tied to an expansion of civil rights and political participation. Both in Prussia after 1806 and in Russia in 1861, the end of serfdom was connected to efforts to create an army of free men supported by a dynamic, mobile society. Like so much else about the modern state, conscription blended emancipation and compulsion, freedom and restraint, empowerment and discipline.

To be effective soldiers, recruits had to learn lessons that had nothing directly to do with combat. For many of them, the army provided the first glimpses of a world beyond their village, a world of clocks and timetables, written rules and standard measurements. In France, officers were encouraged to teach their men skills that would serve them well in civilian life; some garrisons even converted part of their drilling grounds to the cultivation of new kinds of crops. Everywhere recruits learned the national language, the value of soap and hot water, the taste of new foods, the feel of leather shoes, and the attraction of manufactured goods. Colonel F. N. Maude, a British exponent of military service, argued that armies introduced men to practical habits and values necessary for contemporary life: by teaching discipline they were the "school room for the factory," and by instilling a sense of duty they established "the very cornerstone of modern industrial efficiency."

Although every army taught its soldiers — with varying degrees of success — how to live in the modern world, each had its own political curriculum as well.

In France, officers were encouraged to instill in their recruits not only military discipline and soldierly proficiency, but also civic virtues and republican patriotism. After the scandal surrounding the false charges of espionage against Captain Alfred Dreyfus revealed the political unreliability of the officer corps, officer candidates were required to learn appropriate political and social values. In the words of one manual, though the army existed to defend France against its foreign enemies, it also had "the task of instilling respect for the government of the republic and for property." To do this, the army must become "the school where man learns to live in society and where the citizen is formed."

The German army, like the French, was supposed to be a school of political values, although the curriculum was quite different. German recruits learned loyalty to king and country as well as the traditional virtues of piety and deference on which the regime supposedly depended. General von Eichhorn, the commander of the 18th Army Corps, reminded his officers in 1909 that they must mold their men into "loyal subjects." It was certain, Eichhorn continued, that some of the recruits had already been infected by subversive ideas, but since they were young, such political diseases were superficial; military service "must work like a healing spring, and wash the sickness out of their system." The habits of obedience these young men would acquire during their time in the army were "the best possible guarantee against the undue spread of socialistic doctrine."

In Italy, where the problem of regionalism continued to undermine national cohesion, the government insisted that each military unit be drawn from two districts and that it be stationed in a third —despite the obvious difficulties that this created for mobilization. "The army," wrote a leading Italian military thinker in 1873, "is the great crucible in which all provincial elements come to merge in Italian unity."

Around the turn of the century, the ideal of military service as a form of civic education even began to take hold in Britain, where military institutions had a very different social and cultural role than on the continent. Throughout most of British history, being an officer was a socially acceptable occupation, but serving in the ranks was usually limited to the poor and desperate. No respectable family wanted one of its sons to become a soldier, which often meant dangerous, ill-paid service in some far-off corner of the empire. Toward the end of the nineteenth century, the army as an institution

[handwritten marginal note at top: Hæren som social og politisk dannerse istitution]

seems to have become more popular. Rifle associations, paramilitary organizations like the Boy Scouts (formed in 1908 by an army general), and service in the volunteers spread military skills and values. Richard Haldane, whose army reforms required an increased volume of voluntary enlistments, believed that "the spirit of militarism runs fairly high [in the public—i.e., private—schools] and at the universities. What we propose to do in our necessity is to turn to them, and to ask them to help us by putting their militarism to some good purpose." Reflecting on his own experiences at school, George Orwell wrote that "most of the English middle class had been trained for war from the cradle onwards, not technically but morally."

Throughout Europe, the advocates of military service emphasized the role of the army as a school for citizenship. Spencer Wilkinson, the first professor of war at Oxford, wrote in 1910: "To make the citizen a soldier is to give him the sense of duty to the country and consciousness of doing it, which, if spread through the whole population, will convert it into what is required—a nation." To reform an army, in other words, is to create a nation. Two decades earlier, in a very different political environment, Heinrich von Treitschke made much the same point when he argued that "a genuinely national army is the only political institution that brings citizens together as citizens; only in the army do all sons of the fatherland feel united."

The army was exclusively and aggressively male. Military culture, both in the barracks and outside it, was dominated by a kind of masculinity that emphasized physical strength and courage as well as heavy drinking, sexual adventures, and random violence. If, as Treitschke and many others maintained, the soldier was the ideal citizen and the army the school of nationhood, then it was obvious that only males could be fully qualified members of the national community. True citizenship was essentially and irreversibly reserved for the sons of the fatherland, on whose discipline, sense of duty, and national devotion the country depended.

As the mothers and wives of the nation's soldiers, women had important roles to play in the militarized drama of civic life, but these roles were necessarily secondary and supportive. In the room of a Paris municipal hall where couples came for their civil marriage ceremony, a mural captured the normative relationship of men and women. The scene is ancient Gaul, the mythic birthplace of French nationhood; a sturdy, muscular man, mature but still in his prime, is leaving to join a band of warriors armed for battle, while his wife, surrounded by their children, bids him farewell with a gesture of love and encouragement. The nation depended on both women's domestic virtues and men's military prowess, but there is no doubt about the priority of the latter's heroic action over the former's passive support. "It is not in giving life but in risking life that man is raised above the animal," as Simone de Beauvoir once caustically noted. "That is why superiority has been accorded in humanity not to the sex that brings forth life but to that which kills."

Opsumring

The mass reserve army made the state's capacity to wage war a part of everyday life; it made military service an experience shared by millions; and it made the nation's armed forces an inescapable political and social presence. The army was personified not only by professional soldiers in fancy uniforms, but by crowds of nervous young men on their way to basic training, village boys home on leave, and veterans who gathered on Saturday nights to talk about their common military experience. Soldiers were particularly prominent in capital cities and around public buildings, but hundreds of towns had garrisons or depots where reservists were to gather, armories to store equipment, and hospitals to care for aged and infirm soldiers. Like the school, the post office, and the railroad, the army forged powerful links between state and society, center and periphery, national values and local conditions. There was a good deal of truth in the idea that the army was a school for the nation, a place—to borrow the title of Eugen Weber's excellent book—where peasants became Frenchmen (or Germans or Italians).

Although the army resembled a number of the state's integrating institutions, it differed from them in one fundamental respect: armies are about killing and dying. As such, they are the primary expressions of what the German political theorist Carl Schmitt once called the "state's monstrous capacity," its power over life and death. Armies depend on ways of thinking and acting that do not easily fit into modern society. Success in civilian life requires the ability to calculate individual advantage, work hard, sacrifice momentary pleasures for long-term gains. Civilian life assumes stability, order, a predictable future. Armies, on the other hand, must emphasize the importance of group loyalty, unquestioned obedience, and un-calculating courage. They must prepare men for the potentially chaotic, inherently unpredictable world of violence where, in the heat of battle, there is no time to think about the long run. The enormously complex military machines that created and sustained mass reserve armies, therefore, were shadowed by the persistent question of whether their citizen soldiers—the workers, farmers, shopkeepers, and civil servants who filled their ranks—would be willing and able to withstand the rigors of war. Military leaders feared that their soldiers might not be able to shed the habits of civilian life and do what had to be done in defense of the fatherland. But between 1871 and 1914, when most Europeans lived in peace, others hoped that these civilian habits would be too powerful to break and that war would become at last an anachronism.

2

* * *

Pacifism and Militarism

O N AUGUST 24, 1898, when the foreign diplomats assigned to the imperial Russian court gathered for their regular weekly meeting with the foreign minister, Count Muraviev, they were handed a memorandum from Czar Nicholas II that began with an extraordinary assertion: "The preservation of a general peace and a possible reduction in the excessive armaments that now burden every nation are ideals towards which all governments should strive." Over the past twenty years, the czar continued, popular support for international reconciliation has arisen throughout Europe; people everywhere groan under the burden of increasingly dangerous and unbearably expensive armaments. The time has come to liberate the state's resources for peaceful purposes by halting the production of weapons and working to create lasting harmony among nations. To discuss these issues, Nicholas invited governments to a conference that would, he hoped, be "a happy overture to the century ahead."

In European capitals, Nicholas's invitation was warmly welcomed —at least publicly. "This suggestion," Emperor William II of Germany assured his imperial cousin, "once more places in a vivid light the pure and lofty motives with which your counsels are ruled, and will earn you the applause of all peoples." In private, William found nothing to applaud. The idea, he believed, was the product of Russia's financial exhaustion, mixed with Nicholas's own "humanitarian nonsense" as well as "a bit of deviltry." Lord Gough, the British envoy in Berlin, told the German Foreign Office that his govern-

Nicolas Fredsmode

ment was completely in favor of the conference. It was unfortunate, he added with a knowing smile, that parliament was not in session, since it could have passed a resolution opposing war. Of course, Gough went on, we all realize that proclamations about peace and disarmament will have no practical results.

Despite this prevailing skepticism, in the end all of the powers accepted the czar's invitation. After the usual disagreements about who was eligible to attend (Italy insisted that the Vatican not be included, Britain vetoed the Boers, Russia managed to get an invitation for Bulgaria even though it was still formally part of the Ottoman Empire), twenty-six countries eventually sent representatives, including all the European states, great and small, the United States, Japan, several Latin American countries, and the Chinese, Persian, and Ottoman empires. Most delegations were led by a high-ranking diplomat, usually an ambassador. The United States, for instance, selected Andrew Dickson White, the former president of Cornell University, who was then serving as the American representative in Berlin. The most important states also sent military experts, including such prominent figures as Captain Alfred Thayer Mahan, a professor at the U.S. Naval War College and the author of the classic *Influence of Sea Power upon History*, and Sir John Fisher, who would soon become the chief architect of British naval reform and expansion.

The international peace conference officially began on May 18, 1899, as 130 representatives gathered in the "House in the Wood," a handsome seventeenth-century palace near The Hague, to hear the celebratory speeches appropriate for such occasions. Since political issues had been expressly excluded from the agenda and the question of general disarmament had swiftly dropped from sight, the delegates concentrated their attention on three sets of problems, which were assigned to separate committees for consideration: the prohibition of certain kinds of weapons, the elaboration of the rules of war that had been drafted at the Geneva conferences of 1864 and 1868, and the resolution of international disputes through mediation and arbitration.

As we might expect, the military professionals at The Hague vigorously opposed any binding agreement about weapons that could work to the advantage of a less scrupulous enemy. War, they argued, was essentially barbaric and cruel. Why quibble over the details? As

Admiral Fisher put it, a humane war was as impossible to imagine as a humane hell. Invoking his credentials as a historian, Captain Mahan added that weapons had never been rejected simply because they seemed barbaric. Nevertheless, there was agreement among the civilians that the production of some weapons—bombs with suffocating gases and expanding bullets, for example—and delivery systems—balloons and submarines—could and should be outlawed by civilized states, at least for a trial period of five years.

The delegates wrestled with questions about the status of combatants who had been wounded or taken prisoner—issues that a state concerned about the welfare of its citizen soldiers could not ignore. Efforts were also made to protect commercial vessels at sea and limit the naval bombardment of nonmilitary targets on land, both attempts to shield as much of civilian society as possible. The German representatives, remembering their army's difficult struggles with French partisans during the war of 1870–71, insisted on drawing a clear distinction between regular troops and civilians. Only the former were legitimate combatants; anyone else with a weapon should be ruthlessly suppressed, without regard to the laws of war. With considerable difficulty, the delegates managed some compromise resolutions on these issues.

Finally, the Hague Conference dealt with proposals to encourage the peaceful resolution of international disputes, which many delegates regarded as the most promising and consequential item on the agenda. No great power was prepared to accept binding arbitration, which would have compromised its sovereign right of self-defense. But many diplomats acknowledged the utility of an international court in which conflicts might be resolved. Arbitration especially appealed to the Americans, who recognized that, should it fail, their great oceanic glacis would give them plenty of time to prepare a military response. The Germans were suspicious of arbitration for the opposite reason; they feared, not unreasonably, that a potential adversary could use the arbitration process to mask military preparations and thus negate their only comparative advantage, swift mobilization. With some reluctance, Germany joined the other powers in accepting the principle that "in questions of a legal nature, and especially in the interpretation or application of international conventions, arbitration is recognized by the signatory powers as the most

resultate på mg

2

Talking about peace. Delegates to the Hague Conference, May 1899.

effective and at the same time the most equitable means of settling disputes which diplomacy has failed to settle." A court was established. In practice, however, arbitration remained voluntary and limited to matters that did not touch a state's honor or vital interests.

On July 29, the conference concluded by publishing a Final Act, consisting of three conventions (on the pacific settlement of international disputes, the laws of war on land, and the laws of war at sea), three declarations (on the prohibition of the production of certain weapons for five years), and seven resolutions, which expressed the participants' "views" on a variety of issues, including the desirability of holding another conference. Except for a few of the most anodyne resolutions, none of these items were adopted unanimously. Several countries, including the United States, added reservations to those provisions they did sign.

It is easy enough to dismiss the Hague Conference as a gathering of knaves and fools, a farcical prelude to the century's impending tragedies. Three months after the conference ended, the British went to war against the Boers in South Africa; less than five years later two major participants, Russia and Japan, were engaged in hostilities; in 1911, Italy launched an unprovoked attack on the Ottoman Empire, which triggered a series of bloody conflicts in the Balkans that eventually produced the catastrophe of 1914. In the face of the rising tide of violence that would inundate Europe during the first half of the twentieth century, the few weeks at The Hague disappear from view, drowned in the blood of millions. But while the skeptics were right about the conference's practical accomplishments, it is worth remembering that behind all of the self-serving rhetoric, bad faith, and vague compromises was the question that would be posed over and over again in the century that followed: How could the destructive power of war, transformed by democratic politics and technological progress, be restrained?

However skeptical they may have been about its practical value, European governments agreed to attend the Hague Conference because they did not want to appear indifferent to the cause of peace. They recognized that, as Czar Nicholas had declared in his original manifesto, "in the last twenty years aspirations for peace had grown in the consciousness of every civilized state." Even Friedrich von Holstein, a German official with no sympathy for pacifist sentiments, recognized that "the idea of disarmament will not die." And while only a minority of Europeans believed that war could be totally abolished and even fewer favored complete disarmament, many no longer believed that war was a desirable or an inevitable political instrument. Bernhard von Bülow, the German foreign secretary and future chancellor, understood this public mood when he wrote the following instructions to the head of Germany's delegation at The Hague: "At the conclusion of the conference we must be able to demonstrate to German public opinion that we attempted to devote our best efforts to the humanitarian work of the conference, at the same time that we avoided unpractical and dangerous alternatives." Bülow recognized that most Germans, like most other Europeans, wanted two things from their government: peace if possible, military victory if necessary.

In the course of the nineteenth century, the idea that war should and perhaps could be avoided began to undermine the assumption, held by most people throughout human history, that war was as much a part of life as suffering and death. For a few warriors and their advocates, war had traditionally been an opportunity for heroic accomplishment or material gain; but for the majority of humanity, and especially for those unfortunate men and women caught in its path, war meant marauding soldiers, ruined crops, burned villages. The most destructive of the Four Horsemen of the Apocalypse, war often arrived in the company of famine, pestilence, and death. It was not until the eighteenth century that a few philosophers like Immanuel Kant began to argue that war might be banished from civilized society, that it was a pathology to be overcome rather than a scourge to be endured. "War appears to be as old as mankind," Sir Henry Maine wrote in the middle of the nineteenth century, "but peace is a modern invention."

The belief in the possibility of peace came from many dimensions of the modern experience, including the optimistic views of human nature that are usually associated with the Enlightenment. In the nineteenth century, its most fecund and influential source was the extraordinary economic growth that transformed Europe and much of the world. This growth seemed to make war impractical and unnecessary. The French economist Jean-Baptiste Say, for instance, argued that war simply did not pay: "the most fortunate war is a very great misfortune . . . conquest is never worth its cost." Say's British contemporary, the Quaker industrialist John Bright, insisted that "nothing could be so foolish, nothing so mad as a policy of war for a trading nation . . . any peace was better than the most successful war." The world of commerce was a world of individuals and firms competing peacefully for economic advantage, not a world of states struggling for survival. In this commercial realm, the British political economist Richard Cobden wrote, the laws of the market function like the force of gravity in the universe, "drawing men together, thrusting aside antagonisms of race and creed, and language, and uniting us in bonds of eternal peace." As these commercial ties drew people together, the role of states and politics would necessarily diminish. "Free trade," Cobden argued in 1842, "by perfecting intercourse, and securing the dependence of countries one upon the other, must inevitably snatch the power from governments to plunge their people into wars."

Not surprisingly, economically motivated advocates of peace were especially prominent in England, the epicenter of free trade and the

England grand
til at ville free

most civilian society in Europe—and also, not coincidentally, a state protected from aggression by geography and the world's most powerful navy.

The peace movement drew support from many different sources. Some Christians opposed all violence on religious grounds, many socialists regarded international war as another form of exploitation and class conflict, and a variety of humanitarians condemned war because of the suffering it inflicted. This moral revulsion was perhaps best personified by Bertha von Suttner, whose popular novel of 1889, *Lay Down Your Arms*, demonstrated war's cruelty and wastefulness by chronicling its impact on a family of Austrian aristocrats much like her own.

While pacifism—the term itself was first used in 1901—was part of European political discourse throughout the nineteenth century, in the 1890s we can find a new sense of urgency in public discussions about the danger of war and the necessity of peace. Among the most prominent expressions of this mood was a monumental book, *The Future of War*, which first appeared in 1898 in a six-volume, four-thousand-page Russian edition, and was then published in French and German translations as well as in a heavily abridged English version. Its author was Ivan Bloch, who began life as a peddler in Warsaw and had, by the time he was twenty-six, made his reputation (and his first million rubles) building railroads for the imperial Russian government. According to Bloch, Nicholas II, who had become czar in 1894, studied the Russian original of *The Future of War*, summoned him for a discussion of the issues it raised, and then decided to sponsor what would become the Hague Conference of 1899. Bloch died suddenly in December 1901, still deeply engaged in preaching about the futility of war.

Bloch's energy, intellect, wealth, and powerful connections made him a welcome member of the European peace movement, much sought after as speaker, financial supporter, and dinner guest. But he did not share many of his allies' values and assumptions. Bloch insisted that, unlike his friend Suttner, his critique of war was based on scientific analysis rather than sentiment. Unlike religious pacifists, he had no moral objections to war as such; for one thing, he did not think it would be possible to avoid what he called the "frontier brawls" endemic in the Balkans or those minor conflicts in which every empire was engaged.

Fredsbevægelser og nye våben og Bloch

Bloch's point of departure was the characteristically modern conviction that he lived in a time like no other. In the past two decades, he believed, war had been changed beyond recognition, thus
rendering the accumulated weight of historical experience worthless for understanding the nature and implications of armed combat.
Behind this transformation of war was the same process that had
changed so many other aspects of modern life, the application of
technology to the production and distribution of goods. The most
obvious example, Bloch argued, was the breechloading, small-cali-
ber, magazine-fed rifle, which had been adopted by every European
army. These rifles revolutionized combat by greatly increasing the
infantryman's accuracy (up to 2,000 meters) and rate of fire (from
about 1.5 to 4 or 5 rounds per minute). Direct assaults on enemy positions were now nearly impossible. Because the troops' desperate
efforts to escape hostile fire would force them to spread out, the battlefield would expand, thereby making the command and control of
combatants increasingly difficult. Both sides would dig in; the spade
would replace the bayonet as the key to survival. As a result, battles
would end in stalemate while the war itself would drag on without
victory or defeat. No wonder that H. G. Wells, one of Bloch's many
admirers, wrote from the western front in 1916 that he was witness-
ing "Bloch's war."

The great achievement of Bloch's book was not his elaborate
discussion of modern weaponry—much of it commonplace in the
specialized military literature of the day—but rather his powerful
understanding of how these weapons would change war's political
and social meaning. Bloch realized that by demanding a prolonged
and prodigal consumption of resources, weapons production would
put enormous burdens on society as a whole; while its soldiers were
being slaughtered on the battlefield, the civilian population would
confront social disorder, material shortages, and confiscatory taxes.
These privations would make wars more difficult to conclude, since
the winning side would have to demand ever greater compensation
for what it had suffered, whereas among the defeated, "the stoppage
of military operations without attaining the results expected might
easily give rise to revolutionary movements." In the end, the social
fabric would disintegrate, the political order would collapse. The future of war, Bloch wrote, was "not fighting, but famine, not the slay-
ing of men, but the bankruptcy of nations and the break-up of the

whole social organization." <u>Because war would bring nothing but destruction, it was foolish rather than evil</u>—or, it might be better to say, war was evil precisely because it was so foolish.

The leaders of the peace movement regarded the Hague Conference of 1899 as a sign that history was on their side. Baroness von Suttner, for instance, greeted the czar's invitation as a "new star in the cultural heavens." The world, she was convinced, would never be the same again. Suttner, Bloch, and many others showed up at The Hague hoping to make converts among the powerful. Bloch gave a series of four well-attended public lectures on why war no longer made sense; the others mingled as often as possible with the delegates at formal receptions and intimate dinner parties, where they listened eagerly for any sign of hope and encouragement.

Spurred by what they regarded as the growing recognition of modern war's potentially catastrophic impact, members of the European peace movement intensified their efforts in the early twentieth century. <u>In every country pacifists founded or expanded organizations devoted to the promotion of peace. They held conferences and conventions, published periodicals, and sponsored lectures on the evils of war.</u> Two international bodies, the <u>Universal Peace Congress</u> and the <u>Interparliamentary Conference</u>, held annual meetings; after 1892, each maintained a permanent office in Bern, Switzerland. In 1903, the American industrialist Andrew Carnegie endowed a "Peace Palace" in The Hague, which was to house the International Court of Arbitration that had been established in 1899. The palace itself was finished in the summer of 1913, just one year before the Great War began.

Despite intensive efforts at international cooperation, national distinctions remained important within the peace movement, as did philosophical divisions between religious and secular groups, and ideological ones between socialists and liberals. Some peace activists emphasized disarmament because they believed that arms makers encouraged aggression and hostility; others looked to arbitration as a way of settling disputes; some encouraged the development of institutions like the Carnegie Endowment for International Peace, which was founded in 1910; and some, such as the members of the Anglo-German Friendship Society, focused on potential sources of conflict that seemed especially dangerous. This variety reflected the movement's breadth and diverse energy, but it also blunted its impact on decision makers and public opinion.

In 1907, The Hague was the scene of a second international conference on peace. This time the impetus came not from Russia but from the United States, whose original invitation in 1904 had to be shelved until the war that had just broken out between Russia and Japan was over. The second conference was larger (forty-four nations were represented), better prepared, more elaborately organized, and lasted considerably longer than the first. Although the peace lobby was very much in evidence—the British journalist W. T. Stead published a journal devoted to the deliberations—one does not sense the hopeful expectations that had moved people like Bertha von Suttner eight years earlier. Among the participants, mutual mistrust and deep cynicism were even more apparent. Chancellor von Bülow, for instance, instructed his delegates to sign the convention guaranteeing neutral territory in case of war, despite the fact that German war plans called for an invasion of Belgium. "In peace time we should not provoke mistrust," Bülow wrote, "in war we will proceed ruthlessly." As this remark implies, in 1907, as in 1899, governments did not want to appear indifferent to the cause of peace, even if they remained unwilling to compromise their ability to wage war.

In 1910, a new star appeared in the pacifist firmament when an unknown British journalist named Norman Angell published *The Great Illusion: A Study of the Relation of Military Power in Nations to Their Economic and Social Advantage.* It would become one of the early twentieth century's great bestsellers. Frequently reprinted and revised, translated into a dozen languages, Angell's book was widely discussed, praised, and attacked. "Angellism" became a distinct current within the European peace movement. By 1913 there were fifty Norman Angell clubs and ten Norman Angell debating societies in Britain. With the support of some wealthy patrons and a grant of thirty thousand pounds from the Carnegie Endowment, Angell spread his message in a series of lecture tours and a monthly journal, *War and Peace.* Although he lived until 1967, wrote many more books, and took up a variety of causes, *The Great Illusion* remained the foundation of his success, which brought him a knighthood in 1931 and the Nobel Peace Prize two years later. In 1935, the American historian Charles Beard ranked *The Great Illusion* number ten among the twenty-five most important works of the past half century, behind Marx's *Capital* but ahead of Oswald Spengler's *Decline of the West* and John Maynard Keynes's *Economic Consequences of the Peace.*

Born Ralph Norman Angell Lane in Lincolnshire in 1872, Angell grew up in a large, comfortable, but not wealthy provincial family. Considering the single-mindedness that characterized the second half of his life, Lane's early career was oddly restless: he attended several schools, held various jobs, and spent seven years in the United States, including some time in that mecca for the rootless, southern California. After returning to Europe, he worked as a journalist in France and was eventually hired by Lord Northcliffe to run the Paris edition of the *Daily Mail.* In order to keep his book separate from his job as a journalist, when he published *The Great Illusion* Ralph Lane decided to use only his two middle names. So at the age of forty he became Norman Angell. With this new identity he got a new career—he left Northcliffe's employ in 1912 to become a freelance writer and lecturer—and a new mission in life—to persuade Europeans that war could only bring disaster.

Like most influential books, *The Great Illusion* was not particularly original; its success came from Angell's ability to give familiar ideas a distinctive formulation and compelling urgency. Like Bloch, he began with the conviction that conventional wisdom about war and peace no longer fit the contemporary world: "International politics are still dominated by terms applicable to conditions which the process of modern life have altogether abolished." Driving this "process of modern life" was the division of labor—a persistent element in orthodox economics—that was now enhanced by what Angell called a "facility of communication," that expanding network of economic, social, and cultural interaction we now refer to as globalization. By vastly increasing the potential for a division of labor, both domestically and internationally, globalization deepened and intensified economic connections and interdependence. Quantitative changes in long-standing commercial relations were now producing a qualitative transformation of the global economy. All this provided irrefutable proof for Cobden's claim that every nation would lose if the global economy was disrupted by war; the longer the war, the greater the devastation.

[handwritten notes in top margin: "norman Angel lave og han økonomiske argumenter i mod krig"]

[handwritten note in right margin: "god point"]

The expansion of global commerce, Angell argued, had profoundly changed the nature and source of wealth, which no longer depended on the control of territory or resources. In earlier times conquerors occupied valuable land or hauled off objects like gold and precious stones. The shares, contracts, and bonds that represented wealth in the modern era, however, were intangible, their value dependent on a fragile system of institutions whose disruption would injure victors as well as vanquished. A victorious power would not be able to recover the cost of war through territorial annexations because the direct control of a territory produced no economic benefits. Germany, for instance, derived much more economic advantage from its trade with Latin America, where it had no territories, than it did from its expensive conquest of Alsace-Lorraine in the Franco-Prussian War. Monetary indemnities inflicted on defeated states were equally worthless; they simply weakened a potential trading partner without enriching the recipient, as happened when France was forced to pay a substantial indemnity to Germany following its defeat in 1871. It is, therefore, a debilitating illusion to suppose that military power was the necessary foundation for national wealth and well-being—as the example of an economically prosperous but militarily weak state like Switzerland demonstrated. And it was an even more dangerous illusion to think that military victory, no matter how complete and inexpensive, increased the winner's power. For everyone concerned, war was futile because it was incapable of what Angell regarded as the securing of "those moral or material ends which represent the needs of modern civilized people."

Like Bloch, Angell was not a pacifist in the strict sense of the term. "I am not a non-resister," he wrote in 1914. "I believe that aggression should be, and must be resisted, and I would vote any sum necessary to that purpose . . . to the last penny and the last man." He also was a firm believer in the importance of order, which "civilized" states should impose by force if necessary. Conquest for the sake of national aggrandizement was wrong and unproductive, but conquest to create an orderly society was right and progressive: for example, Angell believed that the Germans were wrong to annex Alsace-Lorraine in 1871, but that the United States had been right to annex California. The same line of reasoning led him to argue that Britain should not oppose German efforts to impose order in the Near East. "It is more to the general interest"—of whom, Angell does not say—"to have an orderly and organized Asia Minor under German tutelage than to have an unorganized and disorderly one which should be independent." Armies were necessary and effective when they performed the function of a police force by creating the conditions under which commerce could flourish because this served the interests of the global community, not of a particular state.

Such opinions did not endear Angell to many of his erstwhile allies in the peace movement. E. G. Smith, a longtime opponent of war in all its forms, feared that under Angell's influence, people would "gradually forget the faith and moral enthusiasm which have made the peace movement a live thing." *The Great Illusion* might persuade some people that "war does not pay," but this will not be enough, Smith said, to resist nationalist fevers. "The Angellic vaccination will not save them."

In fact, Bloch and Angell preached a new sort of pacifism, firmly grounded on a realistic assessment of the modern condition. Their ideas had a good deal in common with the traditions of liberal thought, which also emphasized the transformative power of technological innovation and commercial connections. But neither Bloch nor Angell regarded these developments as inherently peaceful; they did not share Kant's conviction that the growth of democracy would create a peaceful world or Richard Cobden's that commerce would render war obsolete. Instead, their hopes for the future were based on fear, fear of modern war's ability to create ever more lethal weapons and to demolish the vulnerable networks on which modern economic life depended. By making people afraid of war's destructive capacity, they hoped to convince them that it was futile. Alfred Nobel, whose invention of dynamite had done so much to make war more destructive, anticipated this argument when he wrote, "I wish I could produce a substance or a machine of such frightful efficiency for wholesale destruction that war would thereby become altogether impossible." Frightfully efficient killing machines shaped the nature of warfare in the twentieth century until July 1945 when, in the high desert near Los Alamos, the explosion of a nuclear device fulfilled Nobel's macabre aspiration.

Few sensible people doubted that a major war between the European powers' huge mass reserve armies would be a perilous business,

Arguments for, worfor wrig angave vor sandsynlig

full of potentially unpleasant surprises. But many also believed that war might be necessary either to pursue some fundamentally important goal or—more likely—to defend the existence of the state. For these people, the arguments of Bloch and Angell were dangerous because they undermined the national will on which victory in a modern war ultimately depended. Historians have not paid much attention to the early-twentieth-century peace movement, but many Europeans at the time viewed it with interest and alarm.

Ivan Bloch's predictions about the future of war, for example, evoked a number of sharp attacks from experts on military affairs. One of Bloch's earliest and harshest critics was Hans Delbrück, a professor of military history at the University of Berlin and the editor of *Preussische Jahrbücher*, an influential political journal. After impatiently dismissing Bloch's scholarly credentials, Delbrück focused on his basic argument that modern weapons had made war too destructive to fight. Why, Delbrück asked, should we believe that? War has always been terrible and yet states have fought each other throughout history. Moreover, if the destructiveness of weapons did make war less likely, then disarmament obviously would not bring peace. The reason why nineteenth-century European powers had been reluctant to go to war was precisely because they were so heavily armed; reducing these arms would simply make war more likely. Delbrück admitted that the modern economies were vulnerable to wartime disruption, but he believed that states would find ways to survive and to turn economic weapons to their own advantage. In any case, fear of economic disruption would not lead people to abandon war. Nor should it. There were still reasons to go to war, among them Germany's struggle for its fair share of the world's resources, which Delbrück regarded as "something worth having even if it had to be purchased with a good deal of blood."

That there were still some things worth fighting for was also the central message of Alfred Thayer Mahan's long critique of Norman Angell, which he published in the *North American Review* in 1912. Then seventy-two and retired from active service, Mahan was still regarded as the world's leading theorist of naval power. Like many military men, he was disturbed by the growing popularity of the movement to eliminate the use of force throughout the world. Pacifism, Mahan believed, was fundamentally wrong: first, because law and order—"from the single policeman to the final court of appeal"—depended on a credible threat of force; second, because sometimes war is necessary and just. Angell's central error was his assumption that nations fought for material gains and that, because war did not pay, it was not worth fighting. In fact, Mahan argued, nations usually went to war for some higher purpose. During the last fifty years, men had fought to end slavery, create nations, free themselves from tyranny, and protect their natural rights. In the face of these higher goals, Angell's shabby calculations of profit and loss were irrelevant. Nor did such calculations count for much in the larger battle to which Mahan himself was deeply committed, the apocalyptic struggle between Christianity and the forces of evil, whose outcome would ultimately be determined by the power of the sword. If the nations of Europe were to disarm, he warned, civilization itself would collapse.

In Britain, where the peace movement was stronger than anywhere else in Europe, pacifism's critics were especially worried about its impact on the national will. British pacifism, these critics maintained, was based on a mistaken sense of invulnerability that encouraged well-meaning Englishmen to become preoccupied with their private lives and to pursue material prosperity at the expense of national security. In order to warn their countrymen against such complacency, a number of British writers created fictional accounts of foreign invasions, a genre that began with Colonel George Tomkyns Chesney's "The Battle of Dorking: Reminiscences of a Volunteer," first published in *Blackwood's Edinburgh Magazine* in 1871, and culminated in a popular play by Saki (H. H. Munro), *When William Came,* in 1913. Chesney's classic established the genre's central message: military defeat was the product of a materialist civilization. His protagonist, an elderly veteran of the battle of Dorking, the fictional site of the decisive German victory, explains to his grandchildren that excessive prosperity, the corrosive effects of individualism, and the false belief that "we were living in a commercial millennium that would never end" had left English society unprepared to resist German aggression. The lesson was clear: "A nation too selfish to defend its liberty could not have been fit to retain it." It was, of course, precisely this selfishness—defined as economic self-interest—that had led Cobden to believe and Angell to hope that war was unnecessary.

[handwritten margin notes: madagunente fil Angels teori / Wells farer]

No one produced a more unnerving account of what a future war might mean than H. G. Wells. Like Chesney and many others, Wells was struck by how modern society's complacency and selfishness made it susceptible to external attack. The plot of *The War of the Worlds*, published in 1898, is set in motion by extraterrestrial invaders, but they remain shadowy figures, without voice or personality. The novel's real subject is not the Martians but the response of British society, which quickly shatters into thousands of isolated, competing groups. When the invaders begin to bomb London, a "roaring wave of fear . . . swept through the greatest city in the world . . . the stream of flight rising swiftly to a torrent, lashing in a foaming tumult round the railway stations, banked up into a horrible struggle about shipping in the Thames, and hurrying by every available channel northward and eastward." By midday, public order had collapsed, producing a "swift liquefaction of the social body." In the end, when the enemy is destroyed by disease, humanity sinks back into the complacency from which it had so rudely been awakened, still unprepared to meet the dangers that surely lie ahead.

The chapter on war in Wells's *Anticipations of the Reaction of Mechanical and Scientific Progress upon Human Life and Thought* (1901) shows the influence of Ivan Bloch's *Future of War*. Like Bloch, Wells emphasized how technology had made war more lethal and decisive victory more difficult to achieve, but he provided a more vivid picture of these new killing machines, including tanks, submarines, and especially airplanes. "No man," he wrote in 1908, "can mark the limits of the destruction of a great European conflict were it to occur at the present time; and the near advent of practicable flying machines opens a new world of frightful possibilities." As Wells was among the first to realize, victory on the battlefield would no longer protect a state from destruction, since airplanes could attack civilian targets and create "unimaginable panics" in the population. Who could predict "what savagery of desperation these new conditions may not release in the soul of man?"

In the light of his ability to imagine what modern war would be like, it should not surprise us that Wells hoped for "universal peace, the merger of national partitions into loyalty to the World State." It is, however, surprising that he viewed the growth of military institutions over the past half century quite positively. Without what Wells called "military urgencies," contemporary society would not be able to overcome its "private self-seeking" and accept the discipline essential for survival. Weapons did not endanger civilization as much as "the undisciplined forces in the collective mind" that might unleash them. In his optimistic moments, Wells foresaw the gradual transformation of military institutions into instruments of progress. Conscription would be a phase through which "the mass of mankind may have to pass, learning something that can be learnt in no other way." Only after people had acquired the "order and discipline, the tradition of service and devotion, of physical fitness, unstinted exertion and universal responsibility" that military institutions provided would universal peace become possible and desirable. Militarism, in other words, was a necessary step toward a world without war.

The American philosopher William James made much the same argument in his remarkable essay "The Moral Equivalent of War," which he first delivered as a lecture at Stanford University in 1906. Although he was opposed to international violence, James nonetheless recognized that war was "the gory nurse that trained societies to cohesiveness." Without it, society would be in danger of losing its vigor, heroism, and capacity for self-sacrifice. Peace, he believed, was not a good thing unless states held on to "some of the old elements of army discipline." This could be done through a system of national service in coal mines and foundries, on fishing boats and construction sites, where young men would learn the toughness and self-control required for individual integrity and social improvement. Only a few years later, ideas like this, stripped of their pacifist garments, would help to persuade intellectuals like John Dewey and Walter Lippmann that intervention in the European war would revitalize American politics and society.

Like many intellectuals around the turn of the century, Wells and James believed that the bonds holding society together were losing their tensile strength; materialism, atomization, and selfishness were becoming the inevitable characteristics of modern life. These fears were given forceful expression in one of the era's most influential books, Gustave LeBon's *La Psychologie des Foules*, first published in 1895, reprinted forty-five times, and translated into seventeen languages, including a widely read English version entitled *The Crowd*. For LeBon, the crowd is a distinctly modern social formation. In many ways it resembles the mob fleeing London in Wells's *War of the Worlds*—fragmented, fractious, easy to manipulate, and prone to panic. Like Wells, who read and admired his work, LeBon argued that military discipline was the best way to overcome the disintegrating forces that produced crowds and thus restore order and cohesion to society. In *La Psychologie Politique et la Défense Sociale*, published in 1910, he advocated the "military spirit" as the only means left to give France "some patience, firmness, and spirit of sacrifice." This "social cement" of patriotic militarism would enable the French to surmount the twin dangers of disintegration and defeat, the insidious influence of pacifism and socialism, and the constant threat of German military power.

Social critics like LeBon usually believed that their own country was particularly susceptible to disorder and disintegration. So it is not surprising that his anxieties were repeated by German theorists, who also feared that modern society—and especially German society—was losing the cohesion necessary for military heroism. Friedrich von Bernhardi, a former Army Corps commander and head of the History Section of the General Staff, noted with dismay that a large part of the German population—and he meant, of course, the urban working class—was no longer committed to national values. Because people lived for the moment, eager to enjoy the pleasures of ordinary life, they had lost sight of the struggle for existence that was, as Heraclitus had written centuries before Darwin, the universal law of life. Wells and William James would certainly not have approved of Bernhardi's glorification of the military aspects of this struggle, but despite differences in tone and intention, their messages were much the same: the danger of war and military discipline make people aware of the importance of a community beyond the life of each individual. All of them would have agreed with Bernhardi's conclusion that military service "not only educates nations in warlike capacity, but it develops intellectual and moral qualities generally for the occupations of peace."

Although Bernhardi's view is often cited as an example of a peculiarly German brand of militarism, there were thinkers like him in every European state. Mahan in the United States, LeBon in France, and many other Europeans would have agreed with the British general Sir Ian Hamilton, who was concerned that, as he wrote in 1905, "up-to-date civilization is becoming less and less capable of conforming to the antique standards of military virtue." Oxford's Spencer Wilkinson was afraid that Britain was losing its capacity to fight. In 1910 he lamented "that we have ceased to be a nation; we have forgotten nationhood, and have become a conglomerate of classes, parties, factions, and sects." His cure for this malady was the familiar call to duty, best expressed in military service, which would restore the nation's sense of itself. On this, Britain's future depended.

In fact, the diagnoses of modern society found in the writings of pacifists and militarists bear a striking resemblance to one another. Bernhardi, for example, was convinced that an "unqualified desire for peace has obtained in our days a quite peculiar power over men's spirits." It has "rendered most civilized nations anemic, and marks a decay of spirit and political courage." Unless this process of decline could be reversed, the nation would stand defenseless before its enemies. Although he spoke about them in different terms, Norman Angell had similar developments in mind when he pointed to the diminishing role of force in contemporary society and wondered whether modern men could keep alive the instincts and abilities that warfare demanded. In his more optimistic moments, Angell suspected that, as a result of these changes, Europeans were "losing the psychological impulse to kill their neighbors." In the ideas of both militarists and pacifists, hopes and fears uneasily coexisted. Militarists hoped that war would restore their nations' collective will; they feared that modern society would not be able to meet the stern test of modern combat. Pacifists hoped that society had outgrown the need for combat; they feared that if war were to come, it would sweep away the fragile structures of civilization.

Pacifism and militarism grew up together; each presupposed and reinforced the other. In societies where war is taken for granted, pacifism is unthinkable; war, like disease and death, may be unfortunate, but it is also an inevitable part of life. In such societies, militarism is unnecessary. It might be desirable to celebrate the warrior's virtues and memorialize his heroic deeds, but there is no need to demonstrate that war itself had a moral purpose or a social function. Only the possibility that war might disappear made it necessary for its advocates to insist that combat was an antidote to the sicknesses of modernity, the best cure for the social debility and cultural decline that infected every nation. The invention of peace, therefore, was inseparable from the invention of a new kind of militarism that concealed behind a dense rhetorical curtain the gruesome realities of modern war.

Pacifism and militarism existed side by side in a Europe that lived in peace but prepared for war. To whom did the future belong? It is not so easy a question to answer as it might at first appear. Many of Bloch's predictions about the destructive capacity of modern weapons turned out to be correct, as did Angell's warnings about the catastrophic social consequences of modern war. Nevertheless, at least in the period between 1914 and 1945, the militarists and their followers took control of powerful institutions, mobilized enormous resources, and demanded discipline and sacrifice from their societies. For thirty years, Europe was shaped by the demands of war. Only after militarism had been thoroughly discredited by the Second World War did the balance tip again, creating a world in which it was the ideas of LeBon and Bernhardi that seemed like artifacts from a long-lost age.

3

Europeans in a Violent World

IN OCTOBER 1913, Baron Günter von Forstner, a twenty-year-
old lieutenant in the 99th Prussian Infantry, was stationed in the
Alsatian town of Zabern. Although quite recently commissioned,
Forstner had already established a reputation for drunkenness and
brutality. When two recruits under his command got into a shoving
match on the rifle range, he told them that if they wanted to fight
they should go into town, adding that if they happened to kill one of
the Wackes—a pejorative term for Alsatians—he would give them
a reward of ten marks. When a local newspaper published what
Forstner had said, the latent hostility between townspeople and the
garrison, Germans and Alsatians, civil and military authorities, came
to a boil. Over the next few weeks Zabern was the scene of several
protests, including stone-throwing by crowds of angry young men.
One evening, the local army commander, who had become impa-
tient with the police's failure to impose order, sent his men to arrest
a number of townspeople, most of them randomly chosen citizens
returning from work. All were released the next morning. Finally,
Lieutenant von Forstner had had enough; when he was heckled by
a few locals, he gave chase, seized a worker on his way to a nearby
shoe factory, and struck him across the head with the blade of his
saber, opening a two-inch wound—the only injury in a month of
demonstrations.

In December, when the German parliament debated the Zabern
affair, an overwhelming majority—including every party except the
Conservatives—condemned the army's behavior and rejected the

zaben episoden
Europa mere sikkert
for indbygger

government's clumsy efforts to defend the military establishment. However, the government survived a vote of no confidence, and not much was done publicly to discipline Forstner and his incompetent superiors. This outcome has led a number of historians, with good reason, to use Zabern as an example of the military's excessive strength and the parliament's regrettable weakness in the prewar German Empire. But there is another, less frequently noted lesson to be drawn from Lieutenant von Forstner's egregious behavior: the widespread outrage and political debates that followed these events also point to the remarkably low tolerance for public violence that existed throughout much of Europe in the early twentieth century. After all, no one died at Zabern, or was even seriously injured. Yet what happened there was enough to cause a political crisis.

lav vold
tolerance

In fact, when compared with any previous period in its history, Europe at the turn of the twentieth century was a peaceful place. Not only had there not been a war between the great powers since 1871; even the specter of violent revolution seemed to be losing its power to frighten or inspire. A number of European radicals, among them Friedrich Engels, now believed that the kind of popular uprising that had toppled the established order in 1789, 1830, and 1848 was a thing of the past. More and more socialists decided that the success of their movement would be gradual and peaceful, the result of political agitation, electoral victories, and trade union activity.

The place of violence in the everyday lives of millions of Europeans significantly diminished in the last decades of the nineteenth century. Street lights, sidewalks, and more effective policing made European cities safer than they had ever been, thereby creating the setting for a new sort of urban sociability. The picnics, family outings, and promenades that are so gloriously depicted in French impressionist paintings would have been unthinkable just a few decades earlier, when Paris was still a dirty and dangerous place. People's appetite for public displays of violence also decreased: executions were moved behind prison walls or, as in France, to an obscure part of the city. The cruel spectacles of bearbaiting and dogfighting slowly disappeared or went underground. Brawling in the streets and brutality in the home, while still all too common, were increasingly viewed as social problems that could be resolved with proper regulations and reforms.

As society became less violent, ordinary men and women came

to feel more secure. In 1899, the British political philosopher Bernard Bosanquet called attention to the significance of this momentous change in the texture of public life: "Hegel observes that a man thinks it a matter of course that he goes back to his house after nightfall in security. He does not reflect what he owes it to. Yet his very naturalness ... of living in a social order is perhaps the most important foundation which the state can furnish to the better life ... Broadly speaking, the members of a civilized community have seen nothing but order in their lives, and could not accommodate their action to anything else."

It was, of course, precisely this accommodation to an orderly life that led some contemporaries to hope, and others to fear, that Europeans had lost not only their taste for violence but also their capacity for heroism. In a series of analyses of modern society, the British philosopher Herbert Spencer noted with satisfaction the decline of violence and the warrior ethos, and their replacement by commerce and calculation, habits of life no longer compatible with war. At the end of the 1890s, the Italian social theorist Gaetano Mosca wrote that if "moral aversions and economic interests" enable the great powers to avoid war for another sixty years, "it is doubtful whether the military and patriotic spirit upon which modern armies are based, and which alone makes possible the enormous material sacrifices that war requires, can be passed on to the rising generation." Mosca, who was anything but an optimist about human nature, had strong misgivings about these changes.

At the beginning of the twentieth century, as at the beginning of the twenty-first, a relatively peaceful Europe lived in a dangerously violent world. A good deal of this global violence was the direct result of Europeans' efforts to extend or maintain control over their imperial dominions. As European missionaries, traders, and settlers penetrated ever more deeply into Asia, Africa, and the Middle East, they required the support of their states, which often found themselves in conflict with local authorities. This was the violent face of the process of globalization that liberal thinkers viewed as the source of international peace and cooperation.

Consider, for instance, what happened just outside the Sudanese town of Omdurman on September 2, 1898, when a British-Egyptian-Sudanese force under the command of General Horatio Her-

bert Kitchener routed an army of sixty thousand followers of the
Mahdi. A local holy man who claimed to have been sent by God,
the Mahdi had organized a movement to drive the Egyptians out
of Sudan, purify their lax version of Islam, and create a theocratic
state under strict Islamic law. In 1883, his soldiers defeated an Egyptian army led by the British; the next year, they captured Khartoum,
across the Nile from Omdurman, and killed the British commander,
General Charles Gordon. The Mahdi himself died in 1885, but
his movement continued under a successor, Khalifa Abdullahi. The
British government sent Kitchener to Sudan with orders to defeat
the Mahdists, avenge Gordon's death, and block French efforts to
establish an outpost on this strategically important sector of the Nile.

Sudan

Omdurman owes its fame to the presence of the young Winston
Churchill, who wrote a vivid account of the battle in *The River War*,
the history of the Sudan campaign he published in 1899. In thirty
pages of sonorous prose, Churchill described the total destruction of
the Mahdists, from the opening salvo of the British artillery at dawn
to the final mopping up by the cavalry, as the shattered army fled
the field a few hours later. Here is his depiction of what happened
when, around midmorning, it appeared that a group of warriors had
isolated the British Camel Corps along the riverbank:

> But at the critical moment the gunboat arrived on the scene and
> began suddenly to blaze and flame from Maxim guns, quick-fir-
> ing guns, and rifles. The range was short; the effect tremendous.
> The terrible machine, floating gracefully on the waters — a beau-
> tiful white devil — wreathed itself in smoke. The river slopes of
> the Kerreri Hills, crowded with the advancing thousands, sprang
> up into clouds of dust and splinters of rock. The charging Der-
> vishes sank down in tangled heaps.

Kitchener's victory was most obviously due to the same magazine-fed rifles, machine guns, and artillery whose revolutionary impact on combat had so impressed Ivan Bloch. When the Khalifa's
troops charged across open ground, they were exposed to the relentless, deadly fire of a virtually invulnerable enemy. At the end of the
battle, 45 British and Egyptian troops had been killed and about 400
were wounded. On the other side, 9,700 died, between 10,000 and
16,000 were wounded, and another 5,000 were taken prisoner.

Less dramatic but no less significant than Kitchener's superior firepower was his long, methodical preparation for the battle. An engineer by training, Kitchener recognized that providing his troops with sufficient food and ammunition was the key to victory. To do this, he had painstakingly constructed a railway across the desert, creating a network of communications in areas where the river could not be navigated. The Khalifa, Churchill concluded, was "conquered on the railway." Without railroads and steamers, colonial troops had to depend on what they could carry or move by horseback across vast and often difficult terrain. As Colonel Charles Callwell wrote in *Small Wars,* one of the earliest and still one of the best accounts of colonial warfare, these wars were "campaigns rather against nature than against hostile armies."

In his introduction to the 1932 edition of *The River War,* Churchill called Omdurman a contribution to "the pacification, restoration, and orderly development of the Sudan." We are more likely to see it as an instance of the violence that persistently attended the expansion and defense of European power over the colonized world. Colonization and war were inseparable. In the eighteenth century, colonial wars had often been the overseas projection of conflicts originating in Europe itself, but after 1815 most colonial wars were between Europeans and indigenous peoples. Some of these were straightforward wars of conquest, but usually they were wars of pacification, the colonizer's violent response to attempts, like the Mahdi's, to shake off foreign rule.

Violence was always close to the surface in colonial encounters because other forms of political persuasion were rarely tried or quickly abandoned. In German Southwest Africa, for instance, the governor's attempts to respond to some of the Africans' demands were frustrated by German settlers and their supporters at home. Here, as was often the case, military conquest and physical annihilation followed in the wake of political failure. This is what Hannah Arendt had in mind when she wrote that "rule by sheer violence comes into play where power [the power to persuade, engage, command] is being lost." Although force is potentially part of all politics, in imperialism it emerges without limit and restraint: "violence administered for power's (and not for law's) sake turns into a destructive principle that will not stop until there is nothing left to violate."

The political and strategic considerations that sometimes mod-

jernbanens betydning
Europas kolonier

ulated conflict between European states were missing in colonial wars; there was no army to defeat in battle, no territory to conquer, no capital to occupy, no government with which a compromise could be negotiated. "In planning a war against an uncivilized nation," wrote Lord Wolseley, an experienced colonial soldier, "your first object should be the capture of whatever they prize most, and the destruction or deprivation of which will probably bring the war most rapidly to a conclusion." What else could this mean but livestock, fields, and villages, perhaps women and children? From the conquest of the Indians on the Great Plains of North America to the slaughter of tribesmen in the southern highlands of Tanganyika, total war—the application of violence without limit and restraint—was inherent in the dynamics of colonial conflict.

In the nineteenth century, by far the most persistently bellicose of the European powers was Great Britain. We usually think of Victorian England as relatively peaceful and accommodating—in contrast, say, to militaristic Germany. In fact, the British army was at war in some part of the world throughout the entire century. While Kitchener was campaigning in Sudan, for example, British troops were fighting on the Northwest Frontier between Afghanistan and India—then, as now, a violent place; in March 1898 they recaptured the strategically vital Khyber Pass that had been lost to Afghan rebels three years earlier. Not long after the Khalifa's death in 1899, Britain joined the other powers in repressing the Boxer uprising in China and went to war in South Africa against the Boers.

Every empire was acquired and maintained with violence. The imposition of French rule over Algeria in the 1830s was a classic case of a war of colonial conquest, which, like many others, evolved into a protracted war of pacification as France tried to extend its control beyond the coast. A latecomer to imperial projects, Germany, in comparison to Britain and France, spilled relatively little blood overseas. But during the first decade of the twentieth century, Germans fought two especially vicious wars in their African colonies. Both were provoked by African resistance to colonial rule, both ended with the devastation of native society.

Even in the grim chronicle of imperial rule, the situation in the Belgian Congo was exceptional in its relentless brutality. In the 1880s, King Leopold of Belgium acquired an immense tract of land in the Congo basin, as large as the United States east of the Mississippi, which he ran as a private corporation. Maintaining order was the job of the Force Publique, composed of African mercenaries with European officers, who used any means necessary to turn the native population into a disciplined workforce for Leopold's rubber plantations. Murder, mutilation, and mass enslavement resulted in a decline in the African population of around ten million. Joseph Conrad, who was there briefly in 1890, used the Congo as the setting for *Heart of Darkness.* It was, he believed, "the vilest scramble for loot that ever disfigured the history of human conscience."

Beginning in the 1890s, a group of reformers organized an international movement to protest atrocities in the Belgian Congo. A self-appointed Commission of Inquiry heard testimony from witnesses. Large public meetings produced petitions and gathered funds for further protests. Parliamentarians sought to enlist their governments to put pressure on King Leopold. The British war against the Boers also became a matter of intense public debate and international criticism. It inspired perhaps the single most influential critique of empire, J. A. Hobson's *Imperialism: A Study,* first published in 1902. The German war of extermination in Southwest Africa was discussed in the Reichstag, where it was supported vehemently by many deputies but condemned by the two largest parties, the Catholic Center and the Social Democrats. Imperialism—the term is a product of the early twentieth century—became an issue to be debated and defended, no longer a fact of life to be taken for granted.

Opposition to imperialism was propelled by many of the same cultural and political currents that animated the peace movement: humanitarian concerns about suffering, a growing reluctance to accept the inevitable place of violence in human affairs, and a deepening skepticism about the alleged economic advantages of the colonial project. When the century began, both movements, the pacifist and the anti-imperialist, enlisted the support of a minority of Europeans. Nowhere were they strong enough to dictate policy. Even the outrage over the Congo, probably the broadest and most active popular movement critical of colonialism, had little impact on what was happening elsewhere in the colonial world.

An important reason why anti-imperialism found so little resonance in the early twentieth century was that the colonial enterprise played a relatively small role in most Europeans' everyday lives. There were, of course, powerful colonial pressure groups in every state. Popular enthusiasm was aroused by colonial victories, just as public outrage usually followed colonial defeats. African explorers like Henry M. Stanley, missionaries like David Livingstone, and soldiers like Gordon of Khartoum were popular figures, widely celebrated in life and elaborately mourned after their deaths. And yet only a small number of people were directly involved in colonial affairs. Some business firms depended on colonial raw materials and hoped — often in vain — to make a profit from colonial investments. A few thousand civil servants spent their lives in colonial administration, and a vast army of Catholic and Protestant missionaries endured great hardships in order to win converts. The overwhelming majority of Europeans experienced their empires indirectly, by contributing money to overseas missions, cheering returning heroes, or vicariously sharing their adventures in pulp fiction or the popular press.

In contrast to the eighteenth century, when European and overseas wars often became entangled, after 1815 the major powers were unwilling to fight one another over an imperial issue. Kitchener slaughtered the Sudanese rebels, but when he met a much smaller French force at Fashoda two weeks later, they exchanged dinner invitations, not gunfire. Colonial rivalries certainly added to the growing tension among European states, especially after 1890; a few large confrontations, such as the one between Germany and France over Morocco, helped to poison the international atmosphere in the years before the Great War. But in one colonial crisis after another, the antagonists pulled back from the brink. The European order was much too valuable and a European war much too dangerous to be risked for some piece of colonial real estate far from home.

No European power wanted to send conscripts to defend its empire — a notable exception to this practice was Italy's campaign in Ethiopia, which ended with the Italian army's disastrous defeat at Adowa in 1896. In addition to professional troops and special colonial armies like the French Foreign Legion, imperial military power depended on native forces, which were led by a fairly small number of European officers and noncoms. Kitchener's 26,000 soldiers at Omdurman, for example, included around 8,000 British regulars and more than 17,000 Egyptian and Sudanese troops. Captain Jean-Baptiste Marchand, the French officer whom Kitchener confronted at Fashoda, commanded 7 French officers and 150 Senegalese light infantrymen. In the Belgian Congo, King Leopold relied on a private army of 6,000, of whom only around 200 were Europeans. Sometimes these native soldiers, like the legendary Ghurkas in India, were from warrior societies; more often, the colonial authorities exploited local rivalries — Eritreans in Ethiopia, Catholic converts in Indochina — to enlist allies. These allies were supposed to do most of the fighting and the dying, thereby keeping the human costs of empire as low as possible. This is why Britain, which was the least militaristic society in Europe, could also be the most warlike in defense of its huge empire. As they carried on their lives in the relatively secure and orderly environment of Victorian Britain, most people accepted without protest the "little wars" their soldiers were waging in the wider world. Because these wars were necessary to defend the national interest, they were part of what it meant to be a state.

On Europe's periphery — along an arc beginning in Ireland in the northwest and extending through the Iberian Peninsula, southern Italy, the Balkans, and on into the vast reaches of imperial Russia — conditions often resembled those in the colonial world: relatively weak state institutions, sometimes in the hands of what was locally regarded as a foreign power; inhospitable and, for the central government, frequently impenetrable terrain; economic underdevelopment; strong communal or kinship loyalties and feeble national commitments. Because the state's authority was weak and its legitimacy contested, political violence was common — both in protest movements against the established order and, more frequently and more lethally, in repressive measures employed by the authorities. As we will see in the next three chapters, imperial violence "came home" after 1914 when, because of the catastrophe of war, more and more of the continent began to resemble its persistently unstable periphery.

Late-nineteenth-century Ireland provides a good example of violence in a quasi-colonial setting. An increasingly aggressive minority of the Irish population had come to view England as an occupying power. At the same time, because economic inequalities overlapped with ethnic and religious divisions, material privation and social antagonism fed, and were in turn nourished by, national conflicts. During the 1870s and 1880s, various forms of political violence, from cattle maiming and the exemplary murder of landlords to the assassination of Lord Frederick Cavendish, the newly appointed chief secretary for Ireland, added to a pervasive atmosphere of menace. After 1900, efforts to craft a political solution to the Irish question were frustrated by the opposition of the Irish Protestants, mostly concentrated in the northern province of Ulster, who realized that independence would leave them a permanent minority in a Catholic state. By 1913, advocates and opponents of home rule had established private armies in both the north and south. It now seemed likely that force would be needed to solve the Irish question. It was, however, by no means certain that the British army would act against the Ulster Volunteers with whom many officers had family and social connections. The outcome of this direct challenge to the state's monopoly of legitimate violence was still undecided when war began in 1914.

Like the Irish, the Spanish had a well-deserved reputation for being difficult to govern. It is no accident that the term "guerrilla" comes to us from the protracted war that Spanish partisans waged against Napoleon. At the end of the nineteenth century, Spain, like Ireland, contained a potent blend of regional, social, and religious animosities, aggravated by the peninsula's tradition of political instability and civil war. Many Basques and Catalans viewed the government in Madrid as an alien, illegitimate authority against which they were prepared to take up arms. In the rural areas of the south and the workers' quarters of major cities, poverty, corruption, and oppression encouraged the growth of anarchist cells dedicated to the use of violence as an instrument of political regeneration.

Between 1859 and 1870, Italian patriots got what Irish, Basque, and Catalan nationalists yearned for, a nation of their own. But this "Italy" was weak and thinly supported; it became a classic example of the gap that often existed between nationalist aspirations and the reality of nationhood. In the decades after 1870, most Italians did not identify with a state that extracted money for taxes and young men for the army without providing much in return. Alienation was especially strong in the south, where the weakness of national institutions created a space into which flowed extralegal intermediaries who provided protection, resolved disputes, and imposed their own kind of order. The best known of these intermediaries was the Sicilian Mafia, which developed a complex and ambiguous working relationship with local elites, whose laws the mafiosi disregarded but whose interests they often served. The core of the Mafia's power was a willingness to kill. In the words of one close observer, "Mafiosi ensure and buttress their intermediate position through the systematic threat and practice of physical violence." Despite the central government's vigorous repressive policies—more Italians were killed in the pacification of the south in the 1890s than in the wars of national unification—the continued presence of groups like the Mafia underscored the state's inability to maintain a monopoly of legitimate force.

Like endemic political violence, the massive social upheavals that had largely disappeared from western Europe remained a potential threat on the periphery. The only major European revolution between 1848 and 1917 occurred in Russia in 1905, when the nation's defeat by the Japanese accelerated deeply rooted forces of unrest and repression. Political crisis at the center of the regime was attended by communal violence and widespread civil unrest; in the immediate aftermath, about 15,000 political opponents of the regime were executed, another 20,000 wounded, and some 45,000 deported to penal colonies. In February and March 1907, a large-scale peasant rebellion swept across Romania, which was finally repressed by the army at the cost of some 11,000 lives. Political violence on this scale had become unthinkable in most of western Europe, where strikes, protests, and demonstrations sometimes led to minor injuries, rarely to fatalities, and never to mass death.

At the beginning of the twentieth century, the most violent place in Europe was the Balkans, where a cluster of new states had managed to wrest independence from the slowly and painfully declining Ottoman Empire. These states were poor, mostly governed by a recently imported dynasty, and divided by a variety of regional, ethnic, and religious antagonisms. All of them nurtured expansionist ambitions directed at establishing—or, as they usually claimed, reestablishing—dominion over national minorities beyond their borders. Serbia, Bulgaria, and Greece, for example, had competing territorial ambitions in Macedonia, a multiethnic province still under the shaky control of the Ottomans. Despite their ambitions, none of these states was able to create a stable and orderly polity at home. In Montenegro, in the harsh mountains where the future Communist leader Milovan Djilas spent his boyhood, family feuds, bloody vendettas, and the constant fear of brigands made violence an ever-present fact of daily life. Djilas called his autobiographical account of these years *Land Without Justice,* which reminds us of Saint Augustine's comment that states without justice were nothing more than large bands of robbers.

In contrast to the political conflicts in Ireland, Spain, or southern Italy, Balkan instability had a direct impact on international order because here local violence intensified the decline of the Ottoman Empire and encouraged the competition of the great powers, especially the two most vulnerable among them, the Austrian and Russian empires. After 1900 these three developments—local instability, Ottoman decline, and great-power rivalry—began to reinforce one another; eventually they would combine to end the long European peace.

A large step on the road to disaster was taken in September 1911 when Italy invaded the Ottoman Empire's North African provinces of Tripoli and Cyrenaica—which, following Roman usage, the Italians would rename Libya. This unprovoked act of aggression illustrates the infectious character of the powers' increasing rapaciousness in the early twentieth century. In the summer of 1911, international attention was focused on a crisis caused by the expansion of French power over Morocco; this crisis not only gave Italy room to maneuver but also evoked the fear that, unless Italy moved quickly, there would be nothing left to take. The result was a classic war of colonial conquest, in which Italian troops rapidly occupied the coast but then found it impossible to pacify the vast interior. After more than a decade of fighting, Italy would control little more territory than it had at the end of the first month.

Although hardly a model of operational skill, the Italian campaign marked one significant innovation in the history of twentieth-century violence: early in the war, the Italian army used airplanes, first to reconnoiter and direct naval gunnery and then to drop grenades on enemy positions. This bombing, the General Staff communiqué noted, had "a wonderful moral effect upon the Arabs." Thus began the airplane's long and bloody career as a weapon of war. It is worth noting that from the beginning the effects of airpower were expressed in the same psychological and moral terms that would be used by its advocates again and again in the decades ahead.

Just as the Italians had been encouraged to act by French aggression in Morocco, so their attack on the beleaguered Ottoman Empire whetted the appetite of the small Balkan states. During the first half of 1912, while the Ottomans were still fighting the Italians in North Africa and the eastern Mediterranean, Bulgaria, Serbia, Greece, and Montenegro put aside their rivalries and agreed to wage a war of aggression aimed at driving the Ottomans out of Europe once and for all. They attacked in October. At first, the Bulgarian and Serbian armies did extremely well; the former pushed the Ottomans to within twenty miles of Constantinople, the latter conquered Kosovo and much of Albania; eventually disease, bad weather, and logistical problems slowed their advance and gave the Ottomans time to recover. By the end of the year the belligerents, now under heavy pressure from the great powers, were ready for an armistice. A peace settlement was signed in London in May 1913. A month later, fighting broke out among the former allies when Bulgaria, dissatisfied with its share of the spoils, attacked Serbian and Greek positions in Macedonia. When the Ottomans and the Romanians joined in, the Bulgarians were forced to seek an end to hostilities, losing in the process most of what they had gained the year before.

The Balkan Wars were both cause and symptom of a deepening crisis in the international system. Distracted by their own animosities, anxieties, and ambitions, the great powers were unable to impose order in this geopolitically critical zone. More and more statesmen —in fragile empires like that of the Habsburgs and in aggressive small states like Serbia—lost faith in the system and began to regard violence as the only reliable defense of their national interests. The arms race, another symptom and cause of international crisis, accelerated at an unprecedented rate. Every state devoted an increasing quantity of men, money, and material to preparing for a war that some decision makers were beginning to regard as unavoidable.

Krig mot osmanerne, The arms race, kommisson for Balkan

The Balkan Wars were also a human disaster of monstrous proportions, foreshadowing the century's worst atrocities. Leon Trotsky, who would soon preside over a civil war in Russia of equal ferocity, was appalled by what he observed during his months as a journalist in the Balkans. Here is how he concluded his report on the situation in Albania at the end of 1912: "Meat is rotting, human flesh as well as the flesh of oxen; villages have become pillars of fire, men are exterminating 'persons not under twelve years of age'; everyone is being brutalized, losing their human aspect. War is revealed as, first and foremost, a vile thing, if you just lift up even one corner of the curtain that hangs in front of deeds of military prowess."

All the participating armies raped and murdered, expelled "enemy" populations from their homes, and wantonly destroyed crops, livestock, and villages. But civilians too were involved, attacking their neighbors in collaboration with friendly armies and taking revenge on the wounded left behind by a retreating enemy. Victims and perpetrators traded roles from one day to the next, chained to one another in a sickening cycle of terror, revenge, and reprisals.

In July 1913, Nicholas Murray Butler, the president of Columbia University and a leading member of the Carnegie Endowment for International Peace, appalled at newspaper reports of the atrocities being committed in the Balkans, decided to act. He asked Elihu Root, a U.S. senator, former cabinet official, and president of the Endowment's board, to empower him to form a commission "to ascertain facts and to fix responsibility for prolonging hostilities and committing outrages." With remarkable speed, an international commission was named and set out for the Balkans just as the second round of warfare was coming to an end. After five weeks of visiting battlefields, gathering evidence, and interviewing witnesses, the commissioners prepared a lengthy report that provides a painfully graphic account of the war's human costs.

The commission's chair, Baron d'Estournelles de Constant, who had been a French delegate at the Hague conferences of 1899 and 1907, lamented, "Every clause of international law relative to war on land and to the treatment of the wounded has been violated by all the belligerents." The most horrifying aspects of the war came from the fact that it "is waged not only by the armies but by the nations themselves . . . This is why these wars are so sanguinary, why they produce so great a loss in men, and end in the annihilation of the population and ruin of whole regions."

Despite the unrelieved carnage that they found in the Balkans, the members of the commission were not pessimistic about the future. Their investigation rested on the conviction that the publication of facts about the war would encourage public support for peace. Moreover, the commissioners took satisfaction from the way the great powers had tried to localize the hostilities. In his introduction to the published version of the commission report, which appeared in the early spring of 1914, d'Estournelles wrote that each of the powers "has discovered the obvious truth that the richest country has the most to lose by war, and each country wishes for peace above all things." About the same time, H. N. Brailsford, a well-informed journalist and one of the two British members of the commission, made the following prediction: "In Europe the epoch of conquest is over and save in the Balkans and perhaps on the fringes of the Austrian and Russian empires, it is as certain as anything in politics that the frontiers of our national states are finally drawn. My own belief is that there will be no more wars among the six great powers."

Even as some Europeans got ready for war, others clung with growing desperation to the hope that they could continue to live in peace.

On June 28, 1914, the heir to the Habsburgs' complex domain, Archduke Franz Ferdinand, was shot to death while visiting Sarajevo, the capital of the Austrian province of Bosnia. His killer was Gavrilo Princip, one of six young terrorists who had been sent to Bosnia by the Black Hand, a shadowy secret society with ties to members of Serbia's security forces. Many Europeans feared that the archduke's assassination would reignite the Balkan conflict; some even worried that it could lead to war among the great powers. But when nothing happened in the days after the assassination, people relaxed. The victim's uncle, Franz Joseph of Austria, remained in his summer residence at Bad Ischl; the German emperor spent most of July on a cruise in the North Sea; the French president and premier departed as scheduled for a leisurely state visit to Russia. Ordinary Europeans heaved a sigh of relief and, like their leaders, went on with their summer plans.

Franz Ferdinand og
etterspillet

But behind the scenes a small group of men were making a series of decisions that would cause one of the greatest catastrophes in modern history. First came the Austrian decision to use Franz Ferdinand's assassination as the occasion for removing the Serbian threat to the empire's interests in the Balkans. Stern action against Serbia, however, raised the danger of intervention by Serbia's patron, Russia. Knowing this, Austria needed to secure the support of its only major ally, Germany. The second bad decision followed on July 5, when Emperor William II responded to Austrian inquiries by impulsively backing any action that Austria thought appropriate; his position was reaffirmed the next day by Chancellor Theobald von Bethmann Hollweg. The Austrians then drafted but did not deliver a harsh ultimatum to Serbia. The Russians, acting on the basis of intercepted messages, warned both Germany and Austria that they would not allow Serbia to be crushed. The French president and premier, by now in St. Petersburg, urged their Russian allies to stand firm: both strategically and politically, France's position in Europe depended on its alliance with Russia. The Austrian ultimatum, with a twenty-four-hour deadline, was finally handed to the Serbian government on July 23. When the Serbians failed to meet all of their demands, the Austrians broke off diplomatic relations and on July 28 declared war; the following day they bombarded Belgrade. At this point, a number of moves ensured that the crisis would escalate to war: Russia began, after some confusion, to mobilize its troops; German efforts to slow things down may or may not have been sincere, but they were certainly ineffective; the French affirmed their support for the Russians; the Russians then ordered a full-scale mobilization on July 30, to which Germany responded, first with an ultimatum, then with a declaration of war against Russia on August 1. The British government, which was divided over whether to go to war, had made desultory attempts to persuade the Austrians to seek mediation but had not made clear its own intentions. After the Germans had violated Belgian neutrality, Britain declared war on Germany on August 4.

Why did the assassination of the unpopular heir to the Habsburg throne bring Europe's long peace to an end? Historians have been debating this question for decades. Their answers tend to fall into one of three broad categories.

[Margin annotations: "1", "Shyld", "2", "De Fischer 'have wager", "Boys will be home before Christmas", "3"]

First, there are those explanations that emphasize the culpability of one of the great powers, usually Germany. This interpretation was inscribed in the Versailles Treaty of 1919, whose Article 231 stated, "The Allied and Associated Governments affirm and Germany accepts the responsibility of Germany and her allies for causing all the loss and damage to which the Allied and Associated Governments and their nationals have been subjected as a consequence of the war imposed upon them by the aggression of Germany and her allies." After falling into disfavor among historians during the 1920s and 1930s, the case for German responsibility was revived in the 1960s by the Hamburg historian Fritz Fischer, and it remains, in somewhat modified form, the most widely held view.

The second explanation is the opposite of the first: the war was no one's, or perhaps everyone's, fault; it was the result of tragic miscalculations, serious blunders, and unfortunate accidents. For understandable reasons, this was the view taken by many of the protagonists when they published their memoirs immediately after the war. The formulation by Britain's wartime prime minister, David Lloyd George, is probably the most famous. "The nations," he wrote, "slithered over the brink into the boiling cauldron of war without any trace of apprehension or dismay."

The third explanation regards the war as the inevitable result of some deeply rooted tensions in the prewar world. For Marxists, the war was sparked by fundamental contradictions in the capitalist economy, for others, by strains in the international system or conservative resistance to pressures for change in the domestic political order. Whatever their source, these tensions, like the seismic energy building between tectonic plates, were bound to erupt in a European war at some point, if not in 1914, then soon thereafter.

It seems to me that none of these explanations—especially as conventionally formulated—is fully convincing, but each one contains an element of truth. That the powers intended to go to war is, in a sense, self-evident: all of them fought because they decided that fighting was better than the alternative, and all were convinced that not going to war would risk their great-power status. And while Fischer's assessment of Germany's motives is unconvincing, there is no question that, by encouraging Austrian aggression against Serbia, Berlin played an essential part in transforming a local conflict into

Grunde til krigen

a European war. But even though the powers chose war, none of them chose the war they got. (Imagine if statesmen in July had been able to see just six months into the future—would any of them have acted as they did?) In this sense, the war was indeed the product of a series of mistakes and miscalculations by all involved. Finally, a war among the great powers may not have been inevitable in 1914, but it would have been difficult to avoid at some point in the early twentieth century. It is hard to imagine how the European international order, which was increasingly being defied by obstreperous minor states, undermined by the aggressive conduct of the great powers themselves, and destabilized by an escalating arms race, could have peacefully absorbed the almost certain demise of the Ottoman Empire and the probable collapse of Austria-Hungary. War did not have to come in 1914, nor did the war have to take the form it did, but some violent clash between the major European states seems highly likely.

One explanation for the outbreak of war in 1914 can easily be dismissed: the war was not a direct response to domestic political crises. There is not a shred of evidence to suggest that, as some historians have argued, statesmen went to war in order to distract their populations from problems at home, avoid political reform, or head off social revolution. Such considerations may help to explain why the war was so hard to stop, but they have nothing to do with why it began. Nor were statesmen pushed into the conflict by nationalist pressure groups or patriotic agitation. However significant popular nationalism might have been as a long-range cause of international tension, it played little or no role in the summer of 1914. At the end of July, when crowds in Berlin waited for news about the Austrian ultimatum, people were anxious and subdued, not belligerent. Most of the popular enthusiasm for the war came immediately after it began, not before.

This is not to suggest that the decision makers in 1914 operated in a political vacuum, without regard to their constituencies. Nowhere did the public push statesmen to act, but everywhere domestic politics set limits on what they could do. No Serbian government, for example, could have accepted the Austrian ultimatum and remained in office—in this regard, the fact that the crisis took place during an election campaign is not irrelevant. The czar's government also

De un neutil at
relge des alliedt

seems to have thought that it could not shirk its responsibilities to Serbia. Russia, Foreign Minister Sergei Sazonov told Nicholas II in the midst of the crisis, will never forgive you if you don't aid the Serbs. In one way or another, many of the participants worried that diplomatic defeat would have unfortunate domestic repercussions. Even when it was not a matter of discussion among the decision makers, such fears certainly played a role in the pressures that made war seem like the least unattractive alternative. The exception seems to have been Britain, where the foreign minister, Sir Edward Grey, was concerned about antiwar sentiment in the public and especially among his own colleagues; this was one reason why he was so reluctant to make the British position on the crisis clear until it was too late to deter Germany's decision to fight.

Every government recognized that it required popular support to fight a modern war: as Helmuth von Moltke, the chief of the Prussian General Staff, wrote to his Austrian counterpart in February 1913, "Your Excellency knows that a war for the very existence of the state requires willing self-sacrifice and enthusiasm from the people." To elicit self-sacrifice and enthusiasm, governments had to convince their populations that they were fighting a defensive war of national survival. The Austrians insisted they were only responding to Serbian aggression; the Germans, that they were defending a loyal ally against attack; the Russians, that they could not see Serbia destroyed; the French, that they could not allow Russia to be overrun by the Germans; and the British, that Belgium had to be defended. These efforts seemed to have worked: just three weeks into the war, Crown Prince Rupprecht of Bavaria wrote in his diary, "Everyone knows what this war, which is forced upon us, is all about; it is a true people's war, whereas if war had resulted from the Moroccan matter, this would not have been understood among the people." Policy makers in Russia, Austria, France, and Britain would have agreed.

Many young men viewed the war as a great adventure, a test of their manhood, a chance for glory and renown. For some, the war offered a release from the routines of everyday life; for others, it was a source of purpose and direction. Rupert Brooke, twenty-seven years old in 1914, handsome, well educated, and enjoying a modest reputation as a poet, remained unsettled and dissatisfied until the war began. Then, in a poem disparagingly entitled "Peace," he gave voice

*Folk positive om krigen —
Alle så det som selvforsvar*

to his delight that the war had delivered "us"—he clearly means his generation—from a "world grown old and cold and weary":

> Now, God be thanked Who has matched us with His hour,
> And caught our youth, and wakened us from sleeping,
> With hand made sure, clear eye, and sharpened power,
> To turn, as swimmers into cleanness leaping . . .

After Brooke died of an infection in the eastern Mediterranean in April 1915, his handful of war poems became famous as lyric celebrations of duty and patriotism, savored by those far from the fighting who could still imagine that in this war "the worst friend and enemy is but Death."

Adolf Hitler, two years younger than Brooke and unlike him in every way save a diffuse dissatisfaction with his place in the world, also greeted the war as a source of meaning. "Overpowered by stormy enthusiasm," he wrote in *Mein Kampf,* "I fell down on my knees and thanked Heaven from an overflowing heart for granting

Welcoming the Great War. Hitler in the crowd, Munich, August 1914.

Hulton-Deutsch Collection/Corbis

me the good fortune of being permitted to live at this time." Hitler was in Munich in August 1914, one step ahead of the Austrian authorities who wanted to know why he had not done his military service. When war was declared, he was part of a jubilant crowd that cheered the news in Munich's Odeonsplatz. A few days later he volunteered for service in the Bavarian army; by October he was in combat on the western front. A competent and courageous soldier, Hitler became a corporal, was wounded, and was decorated with the Iron Cross First Class, an unusual honor for someone of his rank.

In addition to the young and the restless, intellectuals were the war's most conspicuously enthusiastic supporters — or so it seems, since they are the ones who have left the most complete record of their thoughts and feelings. In Vienna, Sigmund Freud applauded Austria's decision to strike against Serbia as "a release of tension through a bold-spirited deed." For a few days he overcame his deep ambivalence about his fatherland: "All my libido," he declared, "is given to Austro-Hungary." Max Weber, a more consistent patriot than Freud, welcomed the war with equal gusto: "No matter what the outcome will be, this war is great and wonderful." In Paris, Émile Durkheim, Weber's only rival as the era's greatest social theorist, believed that the war would contribute to "reviving the sense of community."

How widespread was enthusiasm for the war? The press reported that every major city saw joyous crowds like the one Hitler joined in Munich. Tens of thousands of Berliners gathered at the royal palace to applaud William II, who basked in a rare moment of unqualified popular esteem. In St. Petersburg, where antigovernment demonstrations had taken place just a few weeks earlier, the square in front of the Winter Palace was filled with people singing patriotic songs and praying for victory. Photographs show soldiers marching to their depots surrounded by cheering civilians, who showered them with flowers, chocolates, and good wishes. Troop trains heading west from Germany carried hastily written signs that said *Nach Paris,* while French trains heading in the opposite direction were marked *À Berlin.* None of those young men who look at the camera with such happy confidence would reach their destination; many of them would never see their homes again.

The public's apparent support for war undercut any hope of organized opposition. In Britain, the peace movement was caught by surprise and never succeeded in getting its message across. Norman Angell missed the early stages of the crisis because he had spent part of July in Buckinghamshire leading a workshop on international peace. By the time he returned to London on July 27, Austria's ultimatum to Serbia had already been rejected. Despite his tireless exertions in the few days of peace that remained, Angell was able to do no more than organize a distinguished but isolated group of activists against intervention. On the continent, where the question was not intervention but self-defense, the peace movement was even more diffident and uncertain. Faced with enemy armies massed along their borders, surrounded by patriotic crowds, and aware of the coercive power of the censor and the policeman, few pacifists were able to mobilize against the war; some, like one of the German participants in Angell's Buckinghamshire gathering, rushed off to volunteer for the army.

More surprising and significant than the failure of the peace movement was the political collapse of European socialism, whose millions of supporters were committed to opposing aggressive nationalism and promoting international cooperation. Despite programmatic opposition to militarism, socialist leaders and their constituencies were divided on questions of national defense. No one had a plan for how to prevent an international crisis from developing into war. All denounced wars of aggression, but few were prepared to reject a war in defense of their fatherlands. Unaware of the depth of the crisis until the time for decisive action had passed—as late as July 29, the Executive Committee of the Second International still believed it would be possible to hold an emergency session of the International on August 9—European socialists were paralyzed by ambivalence and division. Victor Adler, a prominent figure in Austrian Social Democracy, saw no possibility of action: "The Party is defenceless ... Demonstrations in support of the war are taking place in the streets ... Our whole organization and press are at risk. We endanger thirty years' work without any political result." In Vienna, as in every other European capital, the socialists ended up supporting the war.

And yet: although the war began without effective opposition, we should not overestimate the breadth and depth of ordinary Europeans' enthusiasm. Not every crowd was festive; in some places the mood was apprehensive. In many cities the streets were empty of cheering throngs, and most rural areas viewed the departure of their young men with profound unease. An American visitor recalled that in the Nuremberg railroad station, "consternation was imprinted on many faces, we saw no elation anywhere, only quiet gloomy resignation." This was the mood that the Berlin journalist Theodor Wolff remembered in July 1916 when he insisted that Germans had not greeted the war joyfully: "Our people had heavy hearts; the possibility of war was a frightening giant nightmare which caused us many sleepless nights. The determination with which we went to war sprang not from joy, but from duty." The article in which these words appear was never published because it did not pass the censor; the newspaper Wolff edited was banned for four months. In Germany, as in the other belligerent states, the government did everything possible to present the picture of a united public, enthusiastically coming to the defense of their nation. The heavier the war's burden and the deeper society's divisions became, the more important it was to preserve this mythic image of solidarity and hope.

The French anthropologist Roger Caillois has compared modern war to ancient festivals in which the individual is uprooted "from his privacy and his personal or familial world in order to be thrown into the whirlpool in which a frenzied multitude noisily affirms its oneness and indivisibility by expending all its wealth and power at one stroke." For some Europeans, the events of 1914 may well have felt like this, at least for a while. But for the majority, the bonds of the private, familial world were not so easily severed. Villagers could not forget that there was hay to stack and fruit to pick, cows to be milked and livestock to be fed—how could these chores be done without the young men who now solemnly left to join their regiments? Parents and lovers knew that some of their hurried farewells would be forever. And in millions of simple homes, families worried about how they would get by without the breadwinner's wages.

Whatever their private anxieties, people followed their leaders into war, some from patriotic conviction, some from ingrained habits of obedience, some because they were relieved to escape their drab routines. Of course they did not know what awaited them, how swiftly and completely the war would change their world, how it would shatter the international order that had kept the peace since 1871, disrupt the economies that had brought unparalleled growth and prosperity, and inject a degree of violence into public life that many Europeans had hoped was gone forever. None of us knows the future, but in 1914 this ignorance has a special poignancy. "Never such innocence again," Philip Larkin wrote in his poem "MCMXIV"— not for those who marched unknowingly into the catastrophe of war, not for future generations who would have to live with the new knowledge of what was now possible.

PART II

A World Made by War, 1914–1945

4

War and Revolution

O N JULY 31, 1914, the graduates of the French military academy at St. Cyr received their commissions. Caught up in the fervor of the moment, Gaston Voizard, one of the freshly minted subalterns, called on his classmates to swear that they would go into battle wearing their dress uniforms, complete with plumed hats and white gloves. Over the next few weeks, the splendidly attired St. Cyrians fell by the score as they led their men against German machine guns; not one member of the class of 1914 would survive the war. A week after Voizard and his comrades swore their oath, Francis and Riverdale Grenfell, officers in His Majesty's 9th Lancers, joined the British Expeditionary Force in Belgium with six horses, two grooms, and their sabers. From an old and distinguished military family (one of their brothers had been killed at Omdurman), the Grenfells had spent most of their lives preparing for the glorious war they now expected to fight. They did not, of course, get the kind of combat for which they were so superbly trained and equipped. Riverdale was killed almost immediately; his twin brother, wounded several times and the recipient of the Victoria Cross, died in Flanders in 1915. "Every war," Paul Fussell reminds us, "is ironic because every war is worse than expected." The Great War is perhaps the most ironic of all because of the terrible distance between expectation and reality, hope and disillusion, sacrifice and achievement.

Europeans went to war in 1914 convinced that their cause was just, that they could win, and that their sacrifices would be worth-

The 9th Lancers at the front, October 1918.

while. "Each side is inspired by its own idealism," H. N. Brailsford, a veteran of the British peace movement, wrote in 1915, "for no government can hope to lead any people into war unless it can convince them that it is fighting under dire necessity for the defense of the right." To keep fighting, governments had to sustain this idealism by proclaiming the war's necessity and justice and by promising that, in the end, victory would be theirs.

"What will the reward of victory have to look like," the German industrialist Walther Rathenau asked after three months of war, "to justify so much blood and tears?" As the flow of blood and tears became a flood, every state felt obliged to promise a future worthy of the sacrifices that victory would require. Statesmen and strategists believed that anything short of victory—that is, any kind of compromise peace without substantial territorial gains—would have dire consequences, certainly for those who made such a peace, and perhaps for the social and political order as a whole. No matter what kind of government they served—republic, constitutional monarchy, autocracy—policy makers were willing to ask their peoples to bear ever greater burdens rather than admit that the burdens they had thus far carried were in vain. This was the vicious circle in which all the belligerents were trapped: the more sacrifices they demanded,

motivation og French plan
XVII

the more essential victory became, which in turn required more sacrifices, and so on and on until one side finally collapsed.

In 1914, there was a broad consensus among professional soldiers that
offensive operations were essential for victory. The French strategist
Jean Colin was convinced that "no one should be allowed to command armies who is not disposed by nature to take the offensive."
In the London *Times,* Charles à Court Repington argued that the
purpose of army engineers was to build roads and bridges, not fortifications. "War is an affair of activity, initiative, and movement."
To think otherwise, he insisted, was to demoralize your troops, encourage the enemy, and open the way to disaster. An army that digs
trenches "is an army that is lost." Courage, energy, discipline—these
were the qualities of mind and spirit that won battles. The technological changes in weaponry that Bloch had written about—and
of which most soldiers were well aware—made these qualities even
more important, as the Japanese had demonstrated in their costly
but decisive victories over Russia in 1905.

Every European army entered the war, therefore, with an offensive
strategy designed to defeat the enemy in great, decisive battles. The
simplest and most straightforward of these was France's Plan XVII,
formulated in 1913, which called for a concentration of French troops
on the eastern frontier. Plan XVII did not specify how these troops
would be deployed, but its underlying principle was clear enough:
"advance with all forces united to attack the German armies."
French mobilization proceeded more smoothly than expected. The
army had anticipated that as many as 13 percent of the troops might
not report to their units; in fact, a negligible number failed to show
up. In the first ten days of August, 4,300 trains carried a million
and a half soldiers to the railheads, from which they then marched
eastward. Dressed in their nineteenth-century uniforms—blue tunics, red breeches, soft hats—the French infantry crossed into enemy territory on August 15, where they met stubborn resistance and
began to suffer appalling losses.

Despite mounting casualties, the French commander, General Joseph Joffre, did everything possible to keep the offensive alive, exer-
cising the kind of leadership that, if it succeeds, seems brave and resolute, and if it fails, inflexible and wasteful. Just as Bloch predicted,

the soldiers' courage could not prevail against modern firepower; within a week, the offensive was broken and the French armies were in retreat, with some units losing up to 80 percent of their effective force. For France, these opening battles were the costliest of the war. On August 22 alone, 27,000 French soldiers were killed, more than 40,000 between August 20 and 23. Three weeks into the war, Plan XVII was in ruins, France's hope for a swift victory extinguished.

While Frenchmen were dying by the thousands on their eastern frontier, the Germans were putting into effect their own offensive strategy, conventionally named after its author, Alfred von Schlieffen, chief of the General Staff from 1891 to 1906. Schlieffen had not produced a detailed operational plan, but he did make clear what was essential for victory: a rapid concentration of forces in the west, which would invade Belgium and northern France and then fight a series of decisive engagements with the enemy. "The French army," he insisted, "must be annihilated." Like the French, the German mobilization went well. Thirty thousand locomotives, 65,000 passenger cars, and 800,000 freight cars moved Germany's 25 army corps, together with their horses, weapons, and supplies, into position. But the further the Germans moved beyond their railheads, the more difficult it became to keep the offensive going. For one thing, Schlieffen's successor, the nephew and namesake of Helmuth von Moltke (the organizer of Prussian victories in the 1860s), was not up to the job. Yet while Moltke made some crucial errors, the basic problem was not leadership but what Clausewitz had called friction—the resistance of circumstances to the commander's plans —which was intensified by the size of the army, the speed with which it was supposed to move, and the distances it had to cover. Moltke could not effectively communicate with or control his forces; he had only vague and incomplete information about the enemy's deployment; at crucial points, essential supplies failed to catch up with his exhausted men.

As the German attack began to lose direction, Joffre managed to recover from the chaotic collapse of Plan XVII. With remarkable calm in the face of potential catastrophe, he was able to put together an effective force from his strategic reserves and the survivors of the failed offensive. Taking advantage of the flank exposed when the German armies turned south, the French engaged them at the

Marne River on September 5. In a bloody four-day battle, the German advance was halted, its momentum broken, the promise of early victory denied. Schlieffen's plan, like Plan XVII, was now inoperable. Over the next several weeks, both sides attempted to regain the initiative, but by mid-December they had clearly stalled. From the Swiss frontier to the English Channel, the French and their British allies faced the Germans along a line of fortified positions that would not move much for almost four years.

Schlieffen had assumed that the Germans would fight a holding action in the east until they had destroyed the French army. But here, too, the unexpected happened: at the end of August, the German 8th Army, under sixty-six-year-old General Paul von Hindenburg, who had been hastily summoned from retirement, soundly defeated a larger Russian force at the battle of Tannenberg. The year's only clear-cut victory, Tannenberg made Hindenburg and his chief of staff, Erich Ludendorff, into national heroes, but it did not win the war in the east. In the three years of combat that followed, the essential difference between the war in the east and the west was spatial: in the west, the fighting was concentrated in a relatively compact area in which the opposing positions were heavily fortified and largely immobile; in the east, where the front ran a thousand miles from Romania to the Baltic, movement and maneuver were still possible. Nevertheless, in the east, as in the west, the opposing armies were mired in a war of attrition that would be decided only when the exhausted Russian army collapsed.

The uniform failure of the opening campaigns destroyed the institutional basis and doctrinal assumptions on which nineteenth-century military thinking had rested. Two months into the fighting, Winston Churchill realized that "this is no ordinary war, but a struggle between nations for life and death." Between the extremes of life and death, survival and extermination, war developed a logic as well as a grammar of its own. Reversing Clausewitz's famous definition of war, politics had become an ancillary of combat, and diplomacy the pursuit of military objectives by other means.

The war's true nature was visible on the western front, where the botched offensives of 1914 were followed by three and a half years of virtual stalemate. Militarily, the reasons for this were simple enough: barbed wire, concrete fortifications, machine guns, and rapid-firing artillery gave troops in defensive positions an overwhelming advantage over those who had to move across open ground to attack them. Looking out at the trenches, Lord Kitchener, the victor of Omdurman and now secretary of state for war, declared: "I don't know what's to be done. This isn't war." But the generals had to do something. This may not have been the war they wanted or expected, but it was the one they had to fight. At first, the military hoped that massive bombardments would be enough to paralyze the defenders by destroying their will to fight, but even when temporarily successful, neither side could turn tactical gains into strategic victory. The defenders were always able to reestablish their positions and counterattack the overextended and sometimes isolated enemy. By the last year of the war, both sides had begun to develop tactics that combined speed, decentralized command, and mobile firepower, but by then Germany had already lost the battle of attrition.

The prolonged agony of the western front echoes in the names of those killing fields that still evoke the war's devastating cost and terrible futility. For the British, the war's exemplary battle was fought along the Somme River in the summer of 1916. Its architect was the new commander of the British forces, Sir Douglas Haig, who was convinced that he had a plan to win the war. During the last week of June, Haig's gunners relentlessly pounded the German positions; on July 1, thirteen British divisions, most of them newly formed units of volunteers who had answered Kitchener's call to arms the year before, emerged from their trenches and began to move toward the German positions. They soon discovered that Haig's predictions were entirely wrong. The enemy's wire remained uncut, his fortifications largely intact, his will to resist unshaken. Of the 100,000 who began the assault, 20,000 were killed, many of them just a few steps from their own trenches, and another 40,000 were wounded; the few British soldiers who reached the German lines were eventually driven back.

What the Somme was for Britain, Verdun was for France and Germany. Like Haig's offensive, the battle for Verdun was conceived as a decisive, war-winning engagement. The German commander, Erich von Falkenhayn, chose the historic fortress of Verdun because he believed the French would have to defend it at all costs; his goal was not to capture the complex of fortifications around the city but

to inflict casualties on the French army. Once France had bled to
death, Britain would have to make peace. The battle lasted from
February to October 1916. Since units were rotated in and out of the
Verdun sector, a large proportion of the French army spent some
time there ("*J'ai fait Verdun,*" the soldiers said. "I've done Verdun.")
To "do" Verdun meant traveling down the road from the town of
Bar le Duc, the "sacred way" along which a French military vehicle
moved every fourteen seconds, twenty-four hours a day. The battlefield presented the new arrivals with a scene of incredible destruction: trees, houses, entire villages had disappeared in a vast waste-
land of mud that was salted with the corpses of men and animals.
At the end of ten months, there were 330,000 German casualties,
379,000 French. Eventually the Germans had to abandon most of
the ground that they had initially captured. The war still had two
more years to go.

Among the numbers that measure the horror of Verdun, perhaps
the most revealing is this: of the 379,000 French casualties, more
than 100,000 were listed as missing. Some of these men may have
wandered off, but the majority had been interred in the mud or simply blown to bits by artillery fire, their bodies unrecovered or un-
recognizable. In the war as a whole, about 300,000 of France's 1.3
million war dead could not be identified. These figures point to the
nature of combat throughout the war, in which a majority, perhaps
as many as 70 percent, of all casualties came from artillery fire; less
than 1 percent were caused by bayonets or sabers, the rest by bullets.
Most of the killing was faceless: soldiers did not see the men they
killed, nor would they see the men who killed them. And the killing
was relentless; unlimited by human stamina, the mechanical delivery
of artillery rounds could go on for hours, even days, as long as there
were shells to feed the guns. At the same time, the guns did their
work with a remorseless contingency: in a quiet section, far from the
front, death could suddenly fall from the sky; in the trenches, some
men died while others, standing next to them, were spared. There
was usually little or nothing a soldier could do to protect himself:
the guns killed without discrimination, the savvy veteran as easily
as the inept recruit. The duration and apparent randomness of artillery fire, together with the feelings of powerlessness it invoked, took
a terrible toll on men's physical and psychological resources. Lord

Moran, who spent two and a half years as a medical officer with the Royal Fusiliers, dryly remarked that "the acid test of a man in the trenches was high explosive; it told each one of us things about ourselves we had not known till then."

In *The Anatomy of Courage*, which Moran wrote during the Second World War while serving as Winston Churchill's personal physician, he defined courage as "will power, whereof no man has an unlimited stock . . . A man's courage is his capital and he is always spending. The call on the bank may be only the daily drain of the front line or it may be a sudden draft which threatens to close the account." Every army contained men who could not or would not keep fighting because their supply of courage was depleted; sometimes they suffered from what was called "shell shock," a condition manifested by a variety of symptoms including extreme agitation, insomnia, delusions, and the apparent loss of speech or hearing. If they were fortunate, shell-shocked soldiers were treated humanely, but usually they were given brief and barbaric therapies intended to force them back to the front as swiftly as possible. In some cases shell shock was regarded as malingering and severely punished, even by a firing squad.

How many soldiers suffered from shell shock? Probably a great many were temporarily affected by a near miss or a particularly intense bombardment, but the number who were totally incapacitated seems to have been quite small. For example, during the battle of Passchendaele, the British army's series of offensives in Flanders from July to November 1917, the authorities reported only 5,346 cases of shell shock out of an army of half a million, so roughly 1 percent. Surely the most remarkable thing about the men who fought in this war was not that some collapsed under the strain, but that most did not.

Nevertheless, 1917 brought signs that the soldiers' ability to absorb pain was reaching its limits. Throughout the year, trouble erupted on every front, worrisome signs that military discipline was beginning to unravel. In May, some French units refused orders to go into the line at the Chemin des Dames; in October, the Italian army at Caporetto collapsed, an entire regiment surrendering to an enterprising young German lieutenant named Erwin Rommel; in November, following the bloody stalemate along the Ypres ridge, came reports

Shell Shock, deception

of increasing insubordination and drunkenness among the British forces. But with the exception of Russia, military effectiveness was restored in every army. Showing a judicious mixture of firmness and conciliation, General Pétain was able to isolate and then suppress the French mutiny; the British reestablished discipline without difficulty; the Italians rallied their forces and managed to stop the German advance. In the twelve months that remained of the war, most soldiers stayed at their posts, obeyed orders, kept fighting despite danger, discomfort, and death.

An element of compulsion is present in even the most highly motivated troops. During the Great War, every army had battlefield police whose job it was to enforce discipline. Harsh punishments, including the death penalty, might be given for sleeping on sentry duty, failing to move forward to engage the enemy, fleeing from danger. On rare occasions, officers simply shot men who did not obey orders; more often, they were brought before military tribunals. The French army executed about 600 of its soldiers during the war, the Italians 750, the British about 400, and the Germans 48. As long as the larger structure of order and discipline remained intact, therefore, the price of disobedience could be very high indeed. The only possible escape from military discipline was desertion, which meant either going over to the enemy (a perilous thing to do under any circumstances, and especially difficult in trench warfare) or attempting to hide somewhere away from the battlefield. A few deserters managed to flee to a neutral country or disappear at home. Many more were caught, tried, and severely punished.

A few men took delight in the perils and possibilities of combat. The German writer and war hero Ernst Jünger, for example, remembers how he went into battle "boiling with mad rage . . . The overwhelming wish to kill gave wings to my feet." Most soldiers, however, fought with grim determination rather than murderous frenzy, doing their duty out of fear, habit, and a commitment to the men closest to them. Patriotic symbols, the rewards of heroism, the gratitude of the fatherland, had more appeal at home than at the front. The great medieval historian Marc Bloch, a junior officer in the French army, did not think that courage relied on national sentiment: "few soldiers, except the most noble and intelligent, think of their country while conducting themselves bravely; they are much

kan sammenlignes med all quiet

more often guided by a sense of personal honor, which is very strong when reinforced by the group."

On the battlefield, the nation seemed far away, and patriotism empty and abstract; what mattered most were the men with whom a soldier fought, on whom he depended, and for whom he would, if necessary, suffer and die. Sometimes the bonds among comrades had first been woven in peacetime associations, regional ties, membership in clubs, even professional organizations. This was the case in the British "pals regiments," which were composed of volunteers who had enlisted together, men from a cricket club or an insurance company, clerks in the same bank or teachers from the same school. But even without a prewar foundation, primary group loyalty could be constructed from the common experience of training and the shared ordeal of the battlefield. One need not love war in order to love one's comrades: there is perhaps no more lyrical account of primary group loyalty than Erich Maria Remarque's *All Quiet on the Western Front,* which describes both the lethal futility of combat and the sustaining beauty of comradeship.

By itself, primary group loyalty did not guarantee an army's effectiveness and fighting power. In order to serve as a source of cohesion reinforcing what Marc Bloch called the soldier's "personal honor," small group loyalties had to be embedded in a larger network of institutions—a unified command structure, a functioning civil society, and a relatively efficient supply system—that provided direction, discipline, and legitimacy, as well as the materials necessary for waging war. If this network ran smoothly, soldiers usually retained the ability and will to fight, even when the larger meaning of the war slipped away. But if the complex bonds linking the army to its society did not hold, then soldiers began to doubt the fairness of their situation, the justice of their cause, the possibility of success. When this happened, the same primary groups that had kept the army together could be transformed into instruments of resistance, mutiny, and disintegration.

Like Norman Angell, General Alfred von Schlieffen recognized that a major war could have devastating consequences for the European economy. This was why Schlieffen based his strategic plan on the assumption that a long war was "impossible at a time when a nation's

Motivation, economic, supplies

existence is founded on the uninterrupted continuance of its trade and industry." Precisely because the war of the future would have to be brief, Schlieffen was convinced that it had to be carefully prepared and quickly won. Most ordinary Europeans also assumed that the war would be over in a few weeks. In 1914, Harold Macmillan, the future British prime minister, was an eighteen-year-old Oxford undergraduate whose main concern was that he would miss the opportunity to be a hero. (Macmillan got his chance. He served with distinction in the Grenadier Guards, almost dying of the wounds he received on the Somme.) In mid-August 1914, two weeks into the war, a British newspaper carried advertisements for the Dresden Conservatory of Music's spring 1915 season; readers were promised more information about ticket availability "on the expectation of a speedy return to normal." In September, the *Economist* explained "the economic and financial impossibility of carrying out hostilities many more months on the present scale."

Since most experts believed that the war was going to be short, they assumed that it would be fought with the men and material at hand. This was why every state had maintained a mass reserve army that was supplied with what seemed like huge stockpiles of equipment. When it became clear that these supplies were being used up much faster than anyone expected, the military did not see what could be done. "Don't bother me with economics," General von Moltke tartly admonished an aide who warned him about dwindling reserves of ammunition, "I am busy conducting the war." Joffre, when told that his soldiers' soft hats should be replaced with steel helmets, replied that they could not be manufactured in time to make any difference.

In his memoirs, Ferdinand Foch, who was among the most eloquent exponents of the offensive élan before the war, reflected on what he had learned during the first months of combat. An army's prospects for victory, he came to realize, were "a function of the engines placed at its disposal. Man alone, however gallant he may be, cannot change them; for without his machines he is powerless." Unrestrained by the frailties of flesh and blood, machines could perform their tasks as long as they were properly supplied with fuel and ammunition. But if the flow of these supplies was interrupted, then these efficient killing machines became useless hunks of metal. After the generals failed to win the battles they had expected to fight in 1914, the machinery of war imposed its own iron logic on the struggle.

The machine that mattered most was the rapid-firing cannon, whose insatiable appetite for ammunition no one had anticipated. When the war began, the German army had 1,000 rounds of ammunition for each cannon. Considering that in the five-month-long war against France in 1870–71 the artillery had fired about 200 rounds per gun, this seemed more than sufficient. The supply lasted six weeks. In France, the situation was no better: the army's store of ammunition was half gone after a month; French cannons fired between 80,000 and 100,000 shells each day, and French factories could produce no more than 12,000. As the war dragged on, the demand for guns and ammunition steadily increased. Between June 24 and July 1, 1916, for instance, 1,400 British guns fired a million and a half shells into the German positions on the Somme — a quarter million in the hour immediately before the attack. In 1918, most guns were firing in one day the number of shells that prewar experts thought would last three or four weeks. By then, French factories were turning out about 1,000 cannons each month, as well as 261,000 artillery rounds and 6 million cartridges each day.

In addition to ammunition, soldiers had to have bread and boots, bandages and blankets, cooking pots and winter coats, candles and tobacco, soap and socks. According to one estimate, every week the German army consumed 60 million tons of bread, 131 million tons of potatoes, and 17 million tons of meat. In the course of the war, British soldiers used 30,000 miles of flannel cloth to clean their rifles; their officers required 50,000 rubber stamps to mark the tons of paper that recorded the constant ebb and flow of combat.

To supply their troops, armies built a dense and extensive system of communications. The British laid almost 7,000 miles of railroad tracks to support their forces in France. Trucks, which had rarely been used for military purposes before 1914, were now essential links between railheads and the front lines, just as cars and motorcycles were used to move men and messages. Airplanes, another recent innovation, became increasingly important to observe enemy positions, support ground forces, and bomb civilian targets. By 1918, British factories were turning out 25,000 planes, the French 24,000, and the amunition, suplies

Germans 17,000. But even in the age of machines, horses continued to bear the burden of human combat, as they had for centuries: in 1917, the British army had 520,000 horses and 230,000 mules on the western front; overall, the British lost about a half million horses during the war.

In 1914, no national economy was prepared for the war it would have to fight; all had to readjust rapidly and radically to unprecedented demands for goods and services. Peacetime production had to be converted to meet military necessities; labor, capital, and raw materials were shifted from firms that made baby carriages and bicycles to those that could turn out machine guns and airplanes. Every economy faced serious problems as it tried to find the resources necessary to feed the war machine. The mass mobilization of young men for the army took them out of the factories and the fields, the disruption of domestic markets and international commerce choked the supply of raw materials, and, most serious of all, the damage caused by foreign occupation or prolonged combat narrowed the economic base. The success of the initial German attack, for instance, deprived France of some of its most valuable industrial plants. On every front, the combatants destroyed scores of villages, ruined mines, and poisoned farmers' fields.

"We have discovered," H. G. Wells wrote during the war, "that the modern economic organization is in itself a fighting machine." The legal institutions and commercial structures that governed the operation of prewar economic life were not sufficient to keep this machine going. Supply and demand, profit and loss, the relatively free circulation of resources, all were incompatible with the unrelenting pressures of wartime. However efficiently markets might distribute resources over the long run, they could not be trusted when the nation's existence hung in the balance; when nothing matters but survival, there is no such thing as the long run. In wartime, a British economist remarked, "the subtleties of comparative advantage become a foolish irrelevance." Like the military's traditional views on the relationship of strategy and statecraft, civilian assumptions about the role of efficient markets and economic calculations were an early casualty of war.

One of the first Europeans to comprehend what the war would mean for national economies was Walther Rathenau, heir to a great industrial empire, amateur philosopher, and fervent German patriot. As early as September 7, 1914, while the battle of the Marne was raging, he wrote to Chancellor Bethmann Hollweg that Germany faced a long and difficult war; he then persuaded the War Ministry to put him in charge of planning for war production. For six months Rathenau and his colleagues laid the basis for Germany's war economy by regulating manufacture, channeling the flow of essential raw materials, and seeking ways to overcome the restrictions imposed by the British naval blockade. Eventually, every belligerent had to create the institutions required to ensure the production of war materials, ration vital commodities, regulate the labor market, and control prices. Each state's response was slightly different—only in Germany, for example, was the military so heavily involved in planning and regulation. No system was without its inefficiencies, profiteering, and sheer waste. And yet all of them, including the Russians, were effective enough to sustain their armies in the field.

In order to maintain what Wells called the "fighting machine," states had to do more than regulate economic activity. The borders over which Europeans had once moved with relative ease were now closely guarded; passports became essential for international travel. Enemy aliens faced internment or expulsion; foreigners were required to have new identity cards. States limited travel, rationed food, and enforced meatless days. To conserve energy, they introduced daylight-saving time. To increase productivity—and decrease the potential for social unrest—they restricted the hours in which pubs could serve beer. By intruding into so many everyday activities, the state established the foundation for regimes of regulation and surveillance that would remain after wartime imperatives ceased. Civilian society would never be the same.

Among the social processes most important to control was the flow of information. No state told its citizens the truth about the war's origins, character, and development. Every government published documents demonstrating that it had been forced to fight. Official reports from the battlefield were often as far removed from reality as the dry and tidy model trenches on display in London's Kensington Gardens. What could be written and reported about the war was tightly controlled by either civilian or military censors. Newspapers that stepped out of bounds might be permanently closed; journal articles and books critical of government policy were banned. In order to prevent the reporting of discouraging news from the front, an elaborate system of military censorship was introduced. In Russia, where surveillance had long been a fact of life, the system was enormously expanded: before 1914, the police had read some 380,000 letters each year; during the war, the censorship commission for one military district opened the same number in three days.

In addition to restricting bad news and critical opinions, every state aggressively sought to strengthen popular commitment to the war. Political authorities had always tried to shape opinion with rituals and ceremonies celebrating power, as well as through schools, election campaigns, and newspapers. The war brought these efforts to new and unprecedented levels—just as the war transformed the state's efforts to regulate social and economic life. Propaganda extolling patriotic virtues and emphasizing enemy vices became an essential part of the war effort in every state.

"War information" offices produced hundreds of thousands of pamphlets and posters, organized rallies and parades, arranged public appearances by war heroes and exhibitions of captured weapons. Film was an especially effective medium to capture the public's imagination; the French government produced more than six hundred newsreels depicting the glorious deeds of its army. Every propaganda campaign attempted to vilify the enemy. The British were particularly skillful in portraying the Germans as bestial—huge, ape-like creatures in spiked helmets, often shown carrying off some innocent maiden's lifeless body. Reports of the atrocities committed by German troops in Belgium during the first weeks of the war were quickly taken up, amplified, and used to demonstrate the instinctive inhumanity of the "Huns." In fact, every side had atrocity stories to tell about its opponents, some—like the terrible tales out of Belgium—based on truth, others carefully fashioned myths designed to dehumanize the foe.

The overwhelming majority of Europe's writers, artists, and scholars eagerly deployed their talent and prestige to denigrate the enemy and defend the justice of the national cause. Their efforts surely influenced some people and helped to confirm the war's legitimacy. But more important than the proclamations of prominent intellectuals was the work of opinion leaders closer to home, the teachers, clergymen, journalists, trade union leaders, and officials to whom people naturally turned for information and direction. This network of local intermediaries helped communities to withstand the strains of war, live with its sorrows and uncertainties, and accept the sacrifices it required.

How well did propaganda really work? After the expectation of rapid victory was disappointed, how long did popular enthusiasm for the war last? In some places and among some groups, enthusiasm never waned. In his compelling account of his boyhood in wartime Germany, the journalist and historian Sebastian Haffner described how he and his friends followed reports from the front, recording the alleged triumphs of their side as though the war were some great and ghastly game. In every nation there were patriots who clung to dreams of victory and celebrated their warriors' heroic deeds. But one suspects that more common in Europe as a whole was the situation in the Isère, a French *département* whose inhabitants remained committed to seeing things through, not from patriotic fervor, but from a grim conviction that the war was just, that it had to be won, and that victory, whatever it might cost, was preferable to defeat.

Like the tough but submissive peasants in the Isère, millions of Europeans continued to support the war. They kept the factories running, harvested the crops, and sent yet another cohort of their young men into battle. With the exception of Russia, widespread support for the war lasted almost until the end; even among the defeated powers, social disintegration came only after the first unmistakable signs of military collapse. The power of the state and the habits of a lifetime combined to keep society together despite the enormous strains that the war placed on the population of every nation.

Among the many illusions destroyed in 1914 was the assumption that in the event of a European war, a clear distinction between soldiers and civilians could be maintained. We saw earlier how the Hague Conference of 1899 discussed ways to protect noncombatants of every sort—not only civilians, but also soldiers wounded in combat and prisoners of war. At the same time, the delegates, and especially the Germans, insisted that only uniformed soldiers had the right to bear arms; because they were neither soldiers nor civilians, irregular

forces were considered bandits, outside the protection of either civil
or military law. Distinguishing between soldiers, who acted as designated agents of the state, and the civilian population was part of a
long series of efforts to limit the use of force in civilized societies; in
peacetime, civil order depended on the state's monopoly of violence;
in war, legitimate violence should be used only by, and against, uniformed military personnel.

   With remarkable speed, the war changed all that. Almost as soon
as the fighting started, the belligerents abandoned the idea that civilian targets were out of bounds. A German zeppelin dropped a
bomb on Paris at the end of August 1914, killing one person. Thereafter both sides periodically attacked each other's cities. In June
1916, for instance, French pilots scored a direct hit on a circus tent
in Karlsruhe, killing 154 German children; in 1917 and 1918, German
planes dropped 300 tons of bombs on London, killing 1,413 people.
Civilians perished when German submarines sank ships headed for
Great Britain—in May 1915, 1,200 people, most of them women
and children, were drowned when the British ocean liner *Lusitania*
was torpedoed. By then, the British had implemented a naval blockade designed to starve the German population; the blockade, which
lasted until the summer of 1919, caused tens of thousands of deaths
and many more serious illnesses. Although no nation was able to inflict the kind of systematic destruction on civilian society that took
place in the Second World War, everyone recognized the advantages
of doing so. "Long distance bombing," wrote the British chief of
staff in January 1918, "will produce its maximum moral effect only
if visits are constantly repeated at short intervals so as to produce in
each area bombed a sustained anxiety." What this officer so politely
referred to as "visits" involved raining tons of explosives on ordinary
men and women as they were going about their daily lives.

   On every front, noncombatants were pulled into the maelstrom
of violence. As millions of men, most of them without combat experience, moved into enemy territory, the potential for error, overreac-
tion, and panic was great. The best-studied case of wartime atrocities concerns the German army in France and Belgium during the
summer of 1914 when the invaders killed more than six thousand civilians, usually in retaliation for largely imaginary attacks by alleged
"partisans." Less well known, but no less murderous, was the behav-

ior of the Habsburg troops in northwestern Serbia, where thousands of civilians, also including women and children, were slaughtered, often with extraordinary cruelty. A report by Professor R. A. Reiss, a Swiss criminologist who investigated these atrocities soon after they occurred, remains one of the most chilling accounts ever written of war's destructive fury. After soberly describing events of unbearable brutality, Reiss concludes that these soldiers, who were psychologically normal, "probably kindly men in private life," initially engaged in acts of "collective sadism" because they were afraid of being massacred themselves; then, "at the first sight of blood," they were transformed into "bloodthirsty brutes" who tortured and killed their victims without cause or hesitation.

Everywhere that a civilian population found itself at the mercy of a foreign army—in eastern Prussia under the Russians, in Alsace under the French, in Belgium, France, and Russia under the Germans—the potential for violence existed. Arson, rape, and murder followed the invaders like scavengers, leaving broken bodies and disrupted lives in their path.

The worst wartime atrocities against civilians were not committed by foreign armies but by agents of the state against their own citizens. In April 1915, the Ottoman government in Constantinople instituted a systematic campaign of annihilation against its Armenian minority. Hundreds of thousands of Armenians were murdered or driven from their homes under conditions of indescribable cruelty and privation. A few found refuge, but most perished at the hands of regular troops and gendarmes, paramilitary formations, bandits, and ordinary citizens. Like everything else about these events, the exact number of victims is contested and necessarily imprecise, but it is clear that by midsummer 1915, virtually all of the empire's ancient Armenian communities had been destroyed.

While the genocide of 1915 was part of a long and complicated historical relationship between Armenians and the Ottomans, it could not have happened outside the context of violence and disruption created by the war. Against these destructive policies, the other European powers were either silent or noisily powerless: the Ottomans' German and Austrian allies were unwilling to intervene, their French, British, and Russian enemies simply unable to. Not for the last time in the twentieth century, the European powers stood by and watched the tragedy of genocide unfold.

civilians, use of
religion

During the first three years of the war, the Russian army displaced as many as six million of its citizens — roughly 5 percent of the population. Some of these people were just caught up in the fighting that raged along the thousand-mile eastern front; others had fled east with the retreating Russian army in order to escape the enemy. Most refugees were forced to move because of official policies designed to cleanse the war zone of "unreliable" groups. Ethnic Germans, even when they had been Russian subjects for generations, were deported to the east, their lands and properties confiscated. In January 1915, the army ordered "all Jews and suspect individuals" to leave the theater of operations. Gypsies, Muslims in the Caucasus, and the inhabitants of the Baltic provinces were also subjected to forced migration. In Russian cities, mobs assaulted Jews and set fire to shops owned by people with German-sounding names.

Governments quickly recognized that they could exploit the ethnic tensions in rival states. The Ottomans, for instance, declared a holy war against their enemies, hoping that this would mobilize the Muslim populations in Russia and in the French and British colonies. With rather more success, the British persuaded Arab leaders to join the battle against the Ottomans, in return for which they were promised national autonomy. In 1917 the British government also called for the creation of "a national home for the Jewish people" in Palestine, which they hoped would rally Jewish support for Britain and its allies. Like their enemies, the Germans played on ethnic divisions. The attempt in April 1916 to spark an Irish rebellion led by Sir Roger Casement ended swiftly when Casement was arrested and hanged as a traitor. The Germans had somewhat more success mobilizing opposition among some of the minorities stretched across Russia's western frontier and in exploiting animosities between the Flemish and the Walloons in Belgium.

The war destroyed the distinction between foreign and domestic politics that had traditionally been a source of order in the European society of states. Every belligerent believed that disrupting the enemy's domestic affairs was a necessary and proper part of the war effort. A potential combatant who could be persuaded to resist his government was one who would not have to be killed. "In the same way as I send shells into the enemy's trenches," wrote the German general Max Hoffmann, "I, as an enemy, have the right to use propaganda against him." Hoffmann was involved in the single most consequential act of enemy interference in another state's internal politics, the decision in March 1917 to transport V. I. Lenin and some thirty other Russian political exiles in a sealed railway carriage from Switzerland back to the motherland, where the Germans hoped they would intensify the political turmoil precipitated by the collapse of the imperial regime.

At the beginning of 1917, Russia confronted an intertwined set of military, social, and political upheavals. Although each crisis had deep historical roots, all were direct and unavoidable products of the war. Poorly equipped and badly led, the Russian army had suffered enormous losses in 1914 and 1915, then recovered and fought with remarkable tenacity—but without achieving a decisive victory. By the end of 1916, morale was low; millions of soldiers yearned for peace. Meanwhile, urban society suffered from the strains of war, the breakdown of infrastructure, and serious disruptions in the food supply. Losses on the battlefield and privations at home fed social discontent and a growing crisis in political confidence. Support for the monarchy was undermined by the czar's incompetence and even more by the rumors of corruption and treachery surrounding his wife and her spiritual adviser, the monk Rasputin. In November 1916, a liberal member of the duma captured the mood of the political elite when he listed the disasters that Russia had recently suffered, punctuating each with the ominous question "Is this stupidity, or is this treason?"

In February 1917, scattered protests in Petrograd quickly escalated into a full-scale rebellion. When the city's garrisons refused to restore order, the centuries-old reign of the Romanovs collapsed, replaced by a provisional government led by moderate reformers. The February revolution's deceptively easy victory was the result of a spontaneous alliance between members of the political elite, who opposed the way the war was being conducted, and millions of ordinary Russians, who opposed the war itself. Differences between these two groups surfaced as soon as the provisional government attempted a new military offensive, postponing vital social and political reforms until after victory had been won. Political demonstrations continued in the cities, peasants sacked manor houses and seized the land, soldiers mutinied, assaulted their officers, fraternized with the enemy, and deserted in growing numbers. "An impassable gulf," one officer wrote, had opened between him and his men. "In their eyes, what has occurred is not a political but a social revolution, which in their opinion they have won and we have lost."

Lenin arrived at Petrograd's Finland Station on April 6. With rare and ruthless consistency, he demanded that his fellow Bolsheviks seize the revolutionary opportunity offered by the war. Unlike the majority of European socialists, including many of his own comrades, Lenin saw no difference between the belligerents; he rejected completely the provisional government's call to defend Russia's newly won freedom against German militarism. But Lenin was no pacifist. Like Clausewitz, whom he had read with great interest, he regarded war as a legitimate political instrument—although for him politics meant the struggle between classes, not states. Since the French Revolution, Lenin believed, war had played a progressive historical role by destroying reactionary elements, weakening repressive institutions, and mobilizing the masses. His goal was to transform the present war into a revolutionary upheaval in which the world's dispossessed classes, including colonial peoples outside Europe, could be mobilized against their imperialist masters. A Bolshevik seizure of power in Russia, which was now possible because of the peculiar circumstances of the war, would help to ignite this global conflagration. Nothing else mattered.

To get his hands on power, Lenin supported popular demands that political authority be given to the Soviets (the councils of workers, peasants, soldiers, and sailors that had spontaneously arisen in February), that the war end immediately, and that land be distributed to the peasantry—despite the strong possibility that these policies would produce political chaos, national humiliation, and social anarchy. At the same time Lenin opened the floodgates of revolutionary change, he used his small but highly disciplined band of followers to brush aside the feeble provisional government and then to hold on to power while he waited for the global revolution to erupt.

From the start, the Bolshevik project was soaked with the blood of its opponents. Driving the violence was Lenin himself: in August 1918, for instance, he instructed his comrades in Penza that they should "hang (by all means hang, so people will see) no fewer than 100 kulaks [peasant landowners], fat cats, bloodsuckers." He sharply rejected any efforts to restrain violence against class enemies: "We are at war to the death. We must spur on the energy and mass character of the terror against the counterrevolutionaries." As the struggle for power in Russia intensified, the spiral of violence increased—not the mechanized, impersonal, random violence of the western front, but direct, face-to-face, immediate brutality inflicted by one person on another, a brutality born of deep-seated class hatred, anger, and aggression but also encouraged and abetted by the new regime.

In the autumn of 1918, Lenin's revolution seemed to be spreading westward, uniting the fragile Bolshevik government with the radicalized masses of central Europe. By the end of September the German High Command realized the war was lost; British and French forces had survived the last desperate German offensive and were now being reinforced by the arrival of more and more fresh American troops. As the generals scrambled to avoid taking responsibility for their impending defeat, the German army's discipline began to unravel and with it the capacity of the German government to maintain order. In the first week of November, while negotiations for an armistice were being conducted, unrest spread across central Europe. Emperor William II fled to Holland, Hindenburg and Ludendorff were relieved of their command, and, apparently following the Russian example, a shaky interim government confronted spontaneously organized councils of soldiers, sailors, and workers. From the Rhine to the Russian frontier, red flags adorned government buildings while armed soldiers and workers patrolled the streets.

Despite the political upheavals that accompanied military defeat throughout central and eastern Europe, the Bolshevik experiment remained confined to Russia. Although the empires of the Hohenzollerns, Habsburgs, and Ottomans, like that of the Romanovs, did not survive the war, the foundational institutions of nineteenth-century states and societies turned out to be much tougher, more cohesive, and more resilient than many had hoped or feared. Considering the suffering the war had inflicted on millions of Europeans, both on the battlefield and at home, it is hardly remarkable that by early 1917 there were outbursts of social unrest and political discontent. The remarkable thing is that they did not come sooner, last longer, and prove more fundamentally disruptive.

Yet while the war did not cause a revolution powerful enough to destroy the European social, economic, and legal order, it did undermine those values and institutions that had shaped European life before 1914. Gone were the habits of cooperation, compromise, and restraint that had helped to regulate conflict among European states between 1815 and 1914. The powerful engines of economic growth that had raised standards of living and encouraged political stability in much of Europe were badly damaged by four years of destruction and dislocation. The state's ability to impose order, and in many places even its monopoly of violence, was badly compromised: riots, political assassinations, and the threat of revolution once again haunted the European imagination. None of these changes prevented people from trying to lead their private lives, looking to their work and their families for meaning and purpose, and hoping for a prosperous present and a secure future. But the pursuit of private happiness had to be carried on in a public world that seemed broken beyond repair.

After 1918, the place of war in European culture was transformed. During the nineteenth century's long peace, most people had accepted the possibility of war even as they lived their lives in the expectation of peace. The mass conscript army, which required that readiness for war be embedded in civilian society, depended on this delicate balance between possibility and expectation, preparing for war and hoping for peace. The war destroyed these delicate balances, opening instead a gulf between civilian and military values and institutions. For millions of Europeans—probably a majority—the war had demonstrated the sad truth of Norman Angell's prediction that war did not pay. Another European conflict, they believed, had to be avoided at all costs. But there were others—a minority, but a militant and active one—who accepted the belief in the regenerating value of violence. In the 1920s and 1930s, therefore, both pacifism and militarism became more robust and politically active, each drawing strength from the other.

5

The Twenty-Year Truce

In the predawn hours of November 11, 1918, the armies confronting one another along the western front sent out their usual raiding parties; in some sectors, the great guns continued to fire until 10:59 a.m. Then, precisely at 11:00, there was silence, as an armistice, the result of difficult negotiations among the belligerents, went into effect. A few soldiers climbed out of their trenches and crossed no man's land to embrace their former enemies; others went off in search of something to drink; many more stayed where they were, still uncertain that the war was actually over. And of course in many parts of the world, the war was not over. Fighting raged for three more years throughout central and eastern Europe and in the Middle East, where armed men struggled to control the shattered fragments of the Hohenzollern, Habsburg, Romanov, and Ottoman empires. Elsewhere the violence died down in 1918, only to ignite again like some imperfectly extinguished forest fire.

The restoration of order to a world shattered by war was the main purpose of the peace conference that met in Paris during the first half of 1919. Although the conference formally included everyone on the winning side, it was dominated by the leaders of the three most important powers, Georges Clemenceau of France, David Lloyd George of Britain, and Woodrow Wilson of the United States. The final settlement, expressed in a series of treaties imposed on the defeated states, was the product of painful compromises among the Big Three, each of whom had brought to Paris very different inter-

Fader, FN

ests, experiences, and political imperatives. As is so often the case, the decision makers compromised at the expense of those who were not at the table. In 1919 this meant, first and foremost, the losers, especially the Germans, the Austrians, and the Hungarians, on whom a harsh peace was imposed, and then a number of the Big Three's less powerful allies, such as the Italians and the Japanese, who did not get the territorial compensation they thought they had been promised.

The great achievement of the conference, certainly flawed in detail but remarkably durable in its broad outline, was a redrawn map of central and eastern Europe that replaced the Habsburg Empire with a group of states roughly based on the principle of self-determination. Considering the divisions among the peacemakers and the magnitude of the task they faced, this was a considerable accomplishment. The settlement's weaknesses came less from the details of its arrangements than from the lack of a concerted will to enforce them. Wilson, who had gotten his way at critical moments in the negotiations, was unable to persuade the U.S. Senate to ratify the treaty. Although heavily involved economically, the United States did not provide political or military support for the peace. Nor did it join the League of Nations, the international organization for which Wilson had had such high expectations. The British signed the treaties, joined the League, but nonetheless began to have misgivings about the peace settlement almost at once. London, for instance, was prepared to guarantee the security of France, but not that of the newly created states in eastern Europe. Under its Bolshevik rulers, Russia, which had been part of the original alliance against Germany, was now outside the society of states. Italy and Japan, although among the victorious powers, were deeply dissatisfied with the status quo. Needless to say, the defeated powers regarded the peace as unfair and basically illegitimate. This left France, much weakened by the human and financial costs of the war, as the only great power that wholeheartedly supported the postwar order.

The most formidable threat to this order was, of course, Germany. Although it had been defeated in battle, its armed forces severely limited, its economy burdened with huge reparations payments to the Allies as well as with the punishing costs of its own war, its government under constant attack, and its society torn by conflict,

Germany remained the most important state in Europe. Indeed, as difficult as it might have been to recognize in 1918, the disappearance of Austria-Hungary and imperial Russia and the losses suffered by France and Italy meant that Germany's *relative* power was greater than it had been when the war began. How Germany would use this power depended on the outcome of a political struggle between the moderate defenders of the new Weimar Republic and their enemies on the extreme right and left. As long as the moderates were in control of the German state, as they were between 1923 and 1930, peace in Europe had a chance. When their control was shaken, as it was between 1919 and 1923, and again between 1930 and 1933, peace was at risk; and when, in 1933, the moderates definitively lost to Hitler's National Socialists, any hope of a European peace was extinguished.

When the war ended in November 1918, most Germans believed they were winning. And why not? They could see that their armies occupied much of eastern Europe and continued to hold positions deep in French territory, at some points only sixty miles from Paris. They did not realize that the failed spring offensives had left the army exhausted, nor could they calculate the moral and material impact of the growing American presence on the western front. When the armistice was signed, Germans' expectations of victory suddenly vanished, and with it the hopes that had sustained them for four years. "We thought we were the Romans," said one contemporary, "when we were actually the Carthaginians." How could this have happened? Many Germans believed that their undefeated armies must have been betrayed, "stabbed in the back," either by the civilian government, to whom the military had deftly shifted responsibility, or by a sinister conspiracy of socialists, Catholics, Jews, and other clandestine enemies of the nation.

The shock of defeat was intensified by the Germans' deep disappointment with the terms of the postwar settlement. During what the prominent theologian Ernst Troeltsch called the "dreamland" between the signing of the armistice and the publication of the peace treaty, Germans could believe that the ideals of democracy and national self-determination, so passionately proclaimed by President Wilson, would enable their new republican government to negotiate a conciliatory peace. These dreams ended in June 1919 when Germany was compelled to accept terms far harsher than any German

Artikel 231 og oprør

had imagined: in addition to ceding territory on every frontier, giving up its colonies and merchant fleet, reducing its army to 100,000 men, and accepting the military occupation of its western provinces, Germany would have to pay its former enemies an unspecified sum in reparations. Germans were particularly outraged by Articles 228 to 231, which demanded that they turn over wartime leaders to be tried as criminals and declared that the war had been caused "by the aggression of Germany and her allies." The overwhelming majority of Germans rejected their sole responsibility for the war—and with it, the moral foundation for the Versailles Treaty as a whole.

In addition to shaping and sustaining a European peace, the great powers had to deal with a series of revolts among their subject peoples throughout the world. There was rioting in British India and demonstrations in favor of independence in Egypt; armed revolts broke out against the French in Syria, the British in Iraq, the Italians in Libya, and the Spanish in Morocco. In Ireland, Britain's oldest and most troublesome colony, a poorly planned and thinly supported armed rebellion had been harshly suppressed in 1916; three years later, Ireland was the scene of a civil war that was finally ended with a treaty granting qualified independence for the largely Catholic south and the continuation of British rule in the Protestant north.

In every case, Europeans responded to resistance with extreme violence. In April 1919, for example, General Reginald Dyer, in order, as he later said, to produce "a moral effect" on the natives, ordered his troops to open fire on a peaceful crowd in the Punjab city of Amritsar, killing 379 and wounding 1,200. To combat Irish nationalists, the British formed a paramilitary group of ex-servicemen, known as the Black and Tans, who responded to terrorist attacks with violent reprisals against the population as a whole. In many parts of the world, the colonial powers used the new military technology to impose their will: airplanes and aerial bombardment proved to be a particularly effective means of surveillance and suppression in the vast spaces of colonial empires. Poison gas, which had been used sparingly in the European war, was dropped from the air against rebellious natives by the British in Afghanistan in 1919 and in Iraq in 1920, by the Italians in Libya in 1923 and 1924 (and then in Ethiopia in 1935, when they would drop five hundred tons of chemical agents), and by the Spanish in Morocco from 1921 to 1927. "I do not understand this squeamishness about the use of gas," declared His Majesty's Secretary of State for War Winston Churchill in 1920. "I am strongly in favor of using poison gas against uncivilized tribes."

In the Middle East, the war was followed by a complex set of armed struggles over the future of what had been the Ottoman Empire. The Ottomans paid dearly for their decision to go to war as Germany's ally: by the time they accepted an armistice in late October 1918, their Arab provinces were in revolt, foreign warships were anchored in the Bosporus, and British troops were preparing to enter Constantinople. The situation further deteriorated when Italy landed forces in Antalya and Greek troops invaded Smyrna and began to move inland. Of these attacks, the last was by far the most perilous, both because of Greece's ambitions to create a modern-day Byzantine Empire and because the Ottomans' Greek subjects joined the invading army in a campaign of communal violence throughout the country.

As his empire rapidly disintegrated, Sultan Mohammed VI, terrified of assassination and politically paralyzed, remained in seclusion in Constantinople. Into this power vacuum stepped Mustafa Kemal, the thirty-eight-year-old commander of imperial forces in Syria who had established his reputation with his successful defense of Gallipoli in 1915. At first Kemal acted in the sultan's name, but increasingly acquired his own sources of legitimacy and authority. By 1923, he had defeated the Greeks, negotiated an end of hostilities with the Allies, signed a new peace treaty at Lausanne, and replaced the sultanate with a republic. In a clear symbolic break with the Ottoman past, Kemal moved the republic's capital to Ankara, a remote Anatolian town far from the mementos of imperial grandeur in Constantinople. The foundation of Kemal's achievement was military victory. "Sovereignty," he told his followers in November 1922, "is acquired by force, power, and by violence."

Force, power, and violence also ensured the survival of Lenin's Bolshevik experiment. By March 1918, Lenin had persuaded his colleagues to accept the harsh peace imposed by the victorious Germans at Brest-Litovsk, through which a third of European Russia, including almost half of its industrial and agricultural capacity, was

kemal og krig i Rusland

lost. Brest-Litovsk, Lenin argued, was necessary for the Bolshevik experiment to survive long enough to merge with the revolution he expected to engulf all of Europe. Once a revolutionary regime was installed in Berlin, a permanent peace settlement would be made by the representatives of the two nations' proletariats. In May, the fragility of the Bolsheviks' authority was revealed when forty thousand Czech prisoners were able to seize control of the Trans-Siberian Railroad. Soon counterrevolutionary armies, with the support of the United States, Japan, Britain, and France, threatened the regime on several fronts. There followed three years of civil war, mass starvation, epidemic disease, and social collapse. That the Bolsheviks emerged victorious was due to the divisions among, and incompetence of, their opponents, as well as to their own ability to create effective military and political institutions based on ideological conviction, patriotic appeals, social animosities, and personal ambition. But most of all, victory resulted from their uninhibited willingness to use force against all enemies, foreign and domestic.

The character of the Soviet Union was shaped by the violence that attended its birth. Among the 6 million combatants who were mobilized for and against the revolution, 1.2 million perished; a half million political opponents of the regime were executed, and another half million people died of starvation. Both the Bolsheviks and their enemies committed atrocities; every side used terror to stifle dissent in the areas it occupied. The Bolsheviks were particularly merciless toward representatives of the old order. The Orthodox clergy, for example, suffered greatly; thousands were murdered, sometimes after being brutally tortured. As many as a quarter million peasants were killed when they refused to hand over their crops or livestock. In some cases, entire communities disappeared. As one eyewitness reported in 1921, "The insurgent Cossack villages have been wiped off the face of the earth ... the men to forced labor in the mines, the women and children scattered everywhere." Leon Trotsky, who had led the Red Army to victory, insisted that it was necessary to "put an end once and for all to the papist-Quaker babble about the sanctity of human life."

By 1921, the Bolsheviks had won the civil war, but the Russian economy was paralyzed, starvation a grim reality, and discontent rampant even among some of the Bolsheviks' hardcore supporters.

Lenin was forced to introduce a series of compromises with peasant producers and small businesses, collectively known as the New Economic Policy. These measures, like the Treaty of Brest-Litovsk three years earlier, were expedients, reluctantly adopted in order to survive. The nature of the Soviet system continued to be determined by techniques of rule forged during the civil war: the domination of the Communist Party over every aspect of social and political life, the merciless use of terror, and the aspiration to create a new kind of society out of the wreckage of Russia's old regime. Lenin, whose physical condition began to deteriorate in 1921, died in 1924 and so did not live to see how his heirs would carry on the prolonged revolutionary war against their own society.

Back in March 1919, at the same time that Lenin was establishing the Third Socialist International (the Second had collapsed in 1915) to spread world revolution from its new center in Moscow, Benito Mussolini and a handful of collaborators founded the Fascio di Combattimento in Milan, the first step along a path that would end three years later when Mussolini became prime minister of the Italian state. Mussolini's rapid rise to power, like Lenin's, was inconceivable without the war. In 1914, the future Duce had been the editor of *Avanti,* the official newspaper of the Italian Socialist Party; soon after the European war began, Mussolini broke with his Socialist comrades over the question of Italian intervention in the war, which he fervently advocated; he served in the army, was wounded, and returned to take his place in the restless and inchoate world of postwar Italian politics. At first Mussolini's program combined radical nationalism with anticapitalist and anticlerical elements from his Socialist past, but he soon moved to the right, praising the monarchy and the church, and promising to defend property and order.

That a man with Mussolini's lack of experience and vague program could become the head of the Italian government reflected the profound social, economic, and political crises in postwar Italy. In 1915, Italy had entered the war on the side of the British, French, and Russians. Instead of the national prestige, territorial expansion, and political renewal the interventionists had promised, the war brought nothing but trouble: Italy's army had been humiliated in battle, its territorial claims frustrated, and its political system strained to the breaking point. In 1919, when millions of disaffected veterans returned home to claim the better lives they had been promised, the economy was in shambles, social unrest widespread, and political instability endemic. With the state nearly paralyzed, many respectable Italians regarded the Fascist squads, recruited from ex-servicemen, students, and young people with a taste for violence, as essential allies against the threat of a Communist-led revolution.

Soviet communism and Italian fascism were, in the words of the French historian François Furet, "the children of World War I." Without the war, Lenin would have remained the exiled leader of a marginal movement, and Mussolini a prominent spokesman for the Italian Socialist Party. But the war not only created the conditions that made the Communist and Fascist regimes possible, it also shaped their character. The Communist and Fascist ideal was a militarized society, organized for war, highly mobilized and tightly controlled, at once democratic and authoritarian, participatory and disciplined.

Violence was centrally important for fascism and communism, not simply as a means of acquiring power but as a transformative instrument, essential to forging a new social and political order. And so both movements brought the war home, domesticated its habits and sensibility, institutionalized its brutality and aggression. "For us," said one Italian Fascist, "the war has never come to an end. We simply replaced external enemies with internal ones." The army, Trotsky proclaimed, "is that school where the party can instill moral hardness, self-sacrifice and discipline." "War," Mussolini wrote in 1932, "brings to its highest tension all human energy and puts the stamp of nobility upon the peoples who have the courage to meet it." The regenerative power of combat, violence, and war, which before 1914 had been celebrated by a few theorists, was now accepted by men in charge of major European states.

Those millions of Europeans who were attracted by fascism or communism (and sometimes by one and then the other) were convinced that prewar values and assumptions were irretrievably lost. The ardent admirers of Lenin and Mussolini no longer believed in the possibility of a liberal order, in a world of commerce and calculation, of peaceful protests and parliamentary debates. Nor could they any longer accept the restraints of traditional strategy and statecraft, the prewar belief in military means justified by political ends, in

clear distinctions between international rivalries and domestic alignments. Those on the extreme left and right agreed that a new order
was needed and that harsh, even brutal measures were required to
achieve it.

Across the lives of those who survived the war stretched the long
shadow of the millions who did not. Of the 70 million men mobilized during the war, 9.45 million were killed. Proportionally, the
smaller, weaker participants suffered most: Serbia lost more than a
third of its army, Turkey and Romania a quarter of theirs, and Bulgaria more than a fifth. Among the major states, France had the
highest percentage of its armed forces killed (16.8), followed by Germany (15.4), Austria-Hungary (12.2), Britain (11.8), and Russia (11.5).
In every army, some units had unusually high casualty rates. In Adolf
Hitler's regiment, the 16th Bavarian Reserve, 3,000 out of a total
of 3,600 men were killed or wounded in less than a month when
they went up against experienced British regulars in the fall of 1914;
within a year, only a handful of the original regiment remained on
active duty. Scotland, which traditionally provided recruits for the
British army's elite regiments, had a fatality rate of 25 percent, more
than twice the average for all of Britain.

In none of the major powers did prominent families try to protect their sons from service at the front. Marshal Foch lost a son
and a son-in-law within a few days of each other in 1914; a month
later, the eldest son of Chancellor Bethmann Hollweg was mortally
wounded on the eastern front; Herbert Asquith, the British prime
minister, never recovered from the death of his beloved son, Raymond, in 1916. It has been estimated that up to one third of the families of British peers lost at least one member. But while death did
not discriminate by wealth or rank or education, it did greatly prefer
the young. The British dead included almost 9 percent of the males
under forty-five. In France, close to a quarter of those conscripted
in the classes of 1911, 1912, and 1913 died in battle; of the 470,000
males born in 1890, half were killed or seriously wounded. One cannot help but wonder how the policy makers could continue to feed
the war machine with the bodies of their own children, but, with
grim determination, feed it they did. "I must be silent here," Bethmann Hollweg wrote to a friend about his son's death. "One man
has no right to complain in view of the hecatombs of the nation."

Measuring the magnitude of these hecatombs helps us to comprehend the war's enormous destructive power, but to understand the intensity of the war's impact on European society we must remember that behind these statistics were millions of individual tragedies. The war left 3 million widows, as many as 10 million orphans, and millions of grieving parents, brothers and sisters, lovers and friends. Virtually every European family included someone who was killed or badly wounded. The survivors would never forget the terrible moment when the news arrived—the phone call or telegram, the grim-faced officer at the door, the village mayor solemnly calling them from the field. Vera Brittain, whose fiancé, brother, and two closest friends were killed, wrote that the war condemned her "to live to the end of my days in a world without confidence and security, a world . . . in which love would seem perpetually threatened by death, and happiness would appear a house without duration, built upon the shifting sands of chance."

Very early in the war, governments recognized that they would have to commemorate the death of each of their citizen soldiers. In 1915, confronted with an unexpected and unprecedented number of fatalities, the French created new military cemeteries and enlarged existing civilian ones. In 1916, the British decided that they needed a separate organization, which eventually became the Imperial (later Commonwealth) War Graves Commission. After the war was over, the French allowed families to move the remains of their loved ones back home: about 300,000 of the 700,000 identified bodies were exhumed and reburied. The British, on the other hand, left their dead in military cemeteries as close as possible to where they had fallen; the graves of more than a million men from throughout the empire, arranged in neat rows with identical white markers and carefully tended by the Graves Commission, were to be found from the Low Countries to the Iraqi desert.

While the democratic nature of modern war made it important to commemorate every soldier's sacrifice, the technology of combat often made it impossible. After every major engagement, the field was littered with the fragmented bodies of the dead, and the killing fields continued to yield a grisly harvest of bones for years thereafter. Throughout the 1920s the British and French armies employed specialists to try to identify newly uncovered skeletons. Despite their best efforts, tens of thousands remained unaccounted for.

In France, the final resting place for many of these men was one of four great ossuaries—at Douaumont near Verdun, Lorette at Pas de Calais, Dorman on the Marne, and Hartmannswillerkopf in Alsace—which were built to house what was left of fallen Frenchmen. These sacred French bones were, the authorities insisted, unmixed with those of their enemies; how exactly French and German remains could be distinguished was not disclosed.

Beginning in London and Paris on November 11, 1920, and then spreading to almost every belligerent nation, a place of honor was reserved for one unknown soldier, whose bones were chosen at random to represent those who died in battle. Located under the Arc de Triomphe in Paris and in the nave of Westminster Abbey in Lon-

The war remembered. Armistice Day, London, 1925.

don, the tomb of the warrior "known only to God" became a ceremonial focus for annual commemorations of the war and the site where visiting dignitaries paid tribute to the nation's war dead. The symbolic value of the unknown soldier was, of course, meaningful only if people assumed that the dead could be identified and their graves could be appropriately marked and preserved. For most of the history of war, only the greatest warriors were remembered; the rest were buried without ceremony in some hastily dug pit. Until the twentieth century, the nameless corpse left on the battlefield was the accepted norm, not the symbolically powerful exception.

The names of the dead were listed on tens of thousands of monuments built throughout Europe—with the notable exception of Russia—after the war. Every one of France's thirty-six thousand communes had a monument recording the sacrifice of each of its fallen sons. Plaques and monuments could also be found in parish churches, college chapels, and places of work; in London's Liverpool Street Station, for instance, a handsome stone tablet lists in eleven long columns the names of the employees of the Great Eastern Railway killed in the war. Such monuments perfectly captured the individual and collective aspirations of the war's commemoration: on most of them, every name is equally important, and all are carved in the same size, usually arranged without regard to rank; at the same time, the list affirms that each individual belongs to a community—a village, neighborhood, parish, school, workplace—whose surviving members must be responsible for remembering his sacrifice. Seen together, these thousands of communities represent the institutional threads from which the nation's common life is woven.

In the 1920s and 1930s, Europeans remembered the war in ceremonies that emphasized sacrifice, grief, and mourning. Soldiers still paraded on national holidays, and guardsmen still stood before public buildings in colorful tunics. Statues of heroes were erected and streets named to recall major victories. But these monuments and rituals were all infused with the knowledge of what war was really like. As in the past, the war's great battles—Somme, Verdun, Passchendaele—evoked memories of courage and survival, but unlike Trafalgar or Waterloo, they also evoked the cost of courage and the suffering that survival had required.

For many Europeans, the private and public legacies of the war were fused, their personal sorrow seamlessly joined to their political frustrations and discontents. The winners briefly hoped that victory would bring them prosperity and security, the "homes fit for heroes" their leaders had promised. "Somebody must suffer for the consequences of the war," the Allied governments told the Germans in June 1919. "Is it to be Germany or only the people she has wronged?" As it turned out, the victors gained little by making the Germans suffer, as was pointed out in John Maynard Keynes's brilliant polemic, *The Economic Consequences of the Peace,* which appeared at the end of 1919. Germany, Keynes argued, would not and could not pay for the war; the harsh peace imposed upon her would create "an inefficient, unemployed, disorganized Europe" in which victor and vanquished alike would suffer. This was, of course, exactly what Norman Angell had predicted would happen, as he tirelessly pointed out, most extensively in *The Fruits of Victory: A Sequel to "The Great Illusion,"* which he published in 1921. However much one might question the specifics of Keynes's and Angell's analysis, the thrust of their argument was hard to deny: the fruits of victory were dry and bitter, the costs of the war persistent and profound.

The mood of disenchantment was captured in a number of remarkable novels and memoirs that began to appear about ten years after the armistice. Most of these books had a common purpose: to reveal what the war was really like by contrasting the experience of combat with what the soldiers themselves had expected, what their leaders claimed, and what many people on the home front continued to believe. Stripped of the rhetoric of heroic action and patriotic sacrifice, the war depicted by these writers is brutal, pointless, and absurd. The soldiers on the other side of no man's land are not the enemy but companions in suffering and death; the real enemy is the war itself, the faceless, relentless, insatiable machine that destroys everything in its path. "The war had won," the British poet Edmund Blunden recalled thinking while at the front, "and would go on winning."

Behind the theme of disenchantment, however, the postwar literature contained another, equally compelling truth about the war: the willingness of citizen soldiers to endure it. Those who wrote most eloquently about the war's horrors—Remarque, Blunden, Robert Graves—also described the patient courage of their comrades.

Would that be possible again? Now that they knew what terrible realities awaited them, would it be possible to send citizen soldiers into battle? After the Somme and Verdun, how could men believe that human will might triumph over the increasingly powerful technologies of death? Did the mass conscript army, on which every European power but Britain had based its national security before 1914, have a future?

Many of the professional soldiers responsible for preparing for the next war thought that the answer to these questions was no. The British, predictably enough, rapidly abandoned the conscription system that they had introduced in 1916. Two years after the armistice their army had been reduced to 300,000 men, scattered across the empire. In contrast to the situation before 1914, British strategists, such as B. H. Liddell Hart and J.F.C. Fuller, stressed the military advantages of highly trained professional troops rather than the moral benefits of conscription. The Germans, too, reduced their army. They had no choice: according to the Treaty of Versailles, Germany could have no more than 100,000 men under arms. While generals and politicians did their best to evade these restrictions, most military planners, like their British counterparts, were skeptical about the value of mass armies. "The whole future of warfare," wrote General Hans von Seeckt, "appears to lie in the employment of mobile armies, relatively small but of high quality." Among the western European powers only France, shadowed by the memories of 1870 and 1914, preserved the mass army, although the length of service was reduced, first to eighteen months and then to a year. Despite the eloquent pleas of military reformers like Charles de Gaulle, the French military budget was cut again and again in the 1930s, leaving the nation increasingly dependent on defensive fortifications constructed along its eastern frontier.

The majority of Europeans fervently hoped that their armies would never be called upon to fight another large-scale war. Peace, most sensible people were convinced, was more than a desirable goal; it was an urgent necessity. The kind of war that Bloch and Angell had predicted was now a shared experience, deeply etched by loss and suffering on the public and private memories of millions of men and women. Pacifism was no longer an eccentric opinion but an unavoidable response to the logic of history. As the British po-

litical theorist R. M. MacIver wrote in 1926, "To assume the closing of the era of national wars is not an act of unscientific utopian-
ism but a reasonable inference from the premise that men in the
long run accommodate their institutions to their necessities." There
was, many Europeans believed, no greater necessity than avoiding
another world war.

MacIver's confidence reflected a new, more positive mood that
began to prevail in European politics in the mid-twenties. The immediate aftermath of the war had been filled with tensions that culminated in 1922, when France and Belgium invaded the Ruhr following Germany's default on reparations payments. By the end of
1923, the new German foreign minister, Gustav Stresemann, initiated a policy designed to obtain concessions from the Allies by
means of conciliation rather than confrontation. Stresemann found
a willing partner in the French premier Aristide Briand, who believed that French security was best defended through closer cooperation with Britain and a more flexible and accommodating policy
toward Germany. The combined efforts of Stresemann and Briand
produced a set of agreements, signed at Locarno in 1925, that guaranteed France's eastern borders. The following year, Germany joined
the League of Nations and participated in a commission to prepare
a general conference on disarmament. Progress was also made on
resolving the vexing matter of reparations, which were finally established at a level the Germans were supposedly willing and able
to pay.

Perhaps the most dramatic expression of this era was the agreement signed in Paris in August 1928 in which the nations of the
world solemnly stated "in the names of their respective peoples that
they condemn recourse to war for the solution of international con-
troversies, and renounce it as an instrument of national policy in
their relations with one another." By declaring that Clausewitz's famous definition of war was illegal and guaranteeing that international disputes would be settled peacefully, this treaty, named for
Briand and the American secretary of state, Frank Kellogg, seemed
to provide the legal foundation for a new international order.

In retrospect, we can see that by 1929 the era of reconciliation in
European affairs was already on its way out. Nevertheless, the optimism it engendered lingered well after the international climate

had begun to change. In September 1930, for example, Nicholas Murray Butler, the longtime president of Columbia University and an enthusiastic promoter of the Kellogg-Briand Pact, evoked the memory of the great nineteenth-century advocate of peace when he offered the "reasonably safe prediction that the next generation will see a constantly increasing respect for Cobden's principles." The following year, Butler was awarded the Nobel Peace Prize. Lord Cecil, speaking to the League of Nations on September 10, 1931, was even more optimistic. "There has scarcely ever been a time in the world's history," he proclaimed, "when war seems less likely than it does at present." The future was not kind to Cecil. Eight days after his speech, the Japanese manufactured an incident at Mukden in order to launch their campaign to conquer Manchuria, beginning a twenty-year spiral of international violence in East Asia. While great-power conflict would not occur in Europe for another eight years, the Japanese aggression in September 1931 represents a plausible starting point for the Second World War.

It is difficult to recapture that evanescent moment when, for Americans like Butler and Europeans like Cecil, the hope for peace burned so brightly. Above all, this hope was nourished by the recognition that the next war would be more terrible than the last. "Who in Europe does not know," Prime Minister Stanley Baldwin said in January 1926, "that one more war in the west and the civilization of the ages will fall with as great a shock as that of Rome?" Even B. H. Liddell Hart, who was very far from being a pacifist, admitted in 1928 that it "is surely clear that any further wars conducted on similar methods must mean the breakdown of European civilization." As it turned out, the widespread fear that war would destroy European civilization was not enough to establish perpetual peace.

Despite Germany's rejection of the peace settlement, a majority of Germans were prepared to support Stresemann's conciliatory diplomacy because they knew what the war had cost them. Like all the belligerents, postwar Germany was awash in grief. During the war, the Germans had mobilized 85 percent of all males between the ages of seventeen and fifty—13.5 million men, more than any other state. Over 2 million of these men had died in battle, another 4.3 million had been wounded. The American historian Robert Whalen has

calculated that in the 1920s almost 10 percent of the total population was composed of either disabled veterans and their families or
the dependent survivors of soldiers killed in action. Of course comparable numbers of people in Britain and France carried the scars of
war on their bodies and spirits; in Germany, the war's victims had to
come to terms with their wounds in a climate of political upheaval,
social discord, and economic dislocation. No European state ben-
efited from its participation in the war. But while the winners may
not have won, the losers certainly did lose — and for them, the collective burdens of defeat added weight to their individual losses and
sacrifice.

Although Germans remembered their dead with thousands of local monuments and millions of private shrines, their efforts to create
national memorials were frustrated by political dissension. In August 1924, for example, a commemoration of the war dead in Berlin ended in fistfights between demonstrators from different parties.
The Stahlhelm, the largest and most active German veterans' association, persistently played on popular dissatisfactions and remained
an often disruptive reminder of the war's unassimilated legacy. Weimar Germany was unable to agree on a national memorial day comparable to November 11 in Britain; instead, Protestants and Catholics, Saxons and Prussians, commemorated the war at different times
and in different ways. Delayed by a variety of regional competitions
and ideological objections, Germany did not commemorate its unknown soldier until 1931, when the Neue Wache, the guardhouse
opposite the royal palace in Berlin, was converted into a tomb of an
unknown soldier, even though no body was ever buried there.

Considering the sea of difficulties they had to navigate, it is remarkable that the moderate defenders of the republic survived the
crises of the immediate postwar period: blame for the defeat, the
need to sign a harsh and to most Germans an illegitimate peace
treaty, the collapse of the monetary system as a result of runaway inflation, a humiliating foreign invasion in 1922, and a series of armed
uprisings from extremists on the right and left. Two things worked
to the republicans' advantage: the courage and skill of a few of its
leaders — especially the president of the republic, Friedrich Ebert,
and the chancellor and then foreign minister, Gustav Stresemann
— and, more important, the fact that, despite widespread hostility

toward the republic, there was no clear alternative to it. In 1922 and 1923, neither the far right nor the far left was strong enough to seize power illegally or win it through free elections.

When the world economy collapsed in 1929, these advantages disappeared. The republic's erstwhile defenders allowed themselves to be outmaneuvered by their conservative opponents, who created the possibility of an antidemocratic alternative. In a desperate effort to harness the Nazi Party's popular support behind their own program, members of the German establishment—quite like their Italian counterparts eleven years earlier—handed the government over to Adolf Hitler in January 1933. More swiftly, ruthlessly, and decisively than Mussolini, Hitler then jettisoned or co-opted his right-wing allies, terrorized his opponents into submission, and dismantled the constitution. In fourteen months he was able to do what the republic had failed to accomplish in fourteen years: establish unchallenged control over the German state and society.

Even more than Lenin and Mussolini, Hitler owed his success to the war. In 1914, he was a man without accomplishments or promise, on the margin of society. Military service gave him a place, a purpose, and his first taste of success. When Hitler's superiors encouraged him to become involved in revolutionary Munich's chaotic political life after the war, he discovered his true calling as an agitator, declaiming against those who had betrayed the fatherland—Communists, republicans, and Jews. From that moment on, the legacy of the war played a critical role at every station along his road to power: in 1923, when the French invasion of the Ruhr tempted him to try to seize power in Bavaria; in 1929, when the campaign against a final settlement of the reparations question led him into a valuable alliance with the conservative nationalists; and in 1933, when he was appointed chancellor by Field Marshal Paul von Hindenburg, the aged war hero who had been elected to his second term as president of the republic the year before.

The war prepared the soil in which Nazism could take root. First, it removed the impediments to extremism that had existed in 1914: a political system based on the rule of law, a dynamic economy, and a widespread patriotic commitment to the state. In their place, the war created sharp political divisions, severe economic dislocation, and deep national frustrations. Second, the war poisoned Germans' civic life, accustomed them to violence, and weakened their commitment to a legal order. Just five years after the political system was thrown into turmoil because of a Prussian officer's mistreatment of civilians in Zabern, Germans were prepared to tolerate extreme violence, including murder, against their political enemies. Without such brutalization, how could millions of respectable people have voted for Hitler and his storm troopers? In a political culture sensitive to the virtues of decency and civility, the Nazi leadership's willingness to condone, even to applaud, the use of lethal force against their opponents — as happened, for example, when drunken storm troopers beat a Communist sympathizer to death in the Silesian village of Potempa in August 1932 — would have disqualified them from serious consideration.

Nazism's appeals were a diverse and often inconsistent collection of fears, hatreds, and special interests. For some Nazis, especially those in Hitler's inner circle, the Jewish question was of central importance; for others, particularly the propertied classes who voted for the Nazis in great numbers after 1930, anticommunism was the key; and for others, especially farmers and small businessmen, the Nazis' hostility to finance capitalism seemed to offer relief from economic ruin. Uniting all these various appeals was the trauma of defeat, which the Nazis insisted had been the work of world Jewry, Communist revolutionaries, and financial manipulators: in 1918, these alien forces had betrayed the fatherland, and they continued to be the source of Germany's discontents. The promise of national rebirth — breaking the shackles of Versailles, restoring national honor, regaining Germany's rightful status as a great power — was at the heart of the Nazi message. It sharply distinguished them from their enemies on the radical left and provided the basis for their cooperation with the respectable right.

Hitler was particularly adept at linking his own experiences to Germans' collective memories of the war. He used the image of the national community created in August 1914, which had been such a decisive personal experience for him, as a model for the fusion of individual and nation. This was the political ideal to which he returned again and again. As Elias Canetti astutely remarked, "His whole subsequent career was devoted to the recreation of this moment . . . Germany was to be again as it was then, conscious of its military striking power and exulting and united in it." Peter Suhrkamp, a war veteran and an opponent of Nazism, recalled that the months after Hitler's coming to power in 1933 reminded him of the first weeks of the war. Suhrkamp was struck by how the crowds of uniformed men, the marching columns of soldiers and storm troopers, the mood of determination, and the shared sense of national purpose all evoked the Germans' initial enthusiasm for battle. "Without the past war, such a pure military phenomenon would hardly have been possible."

Military service was an essential part of Nazism, both as movement and regime. The storm troopers in their brown uniforms, led by veterans and organized along military lines, revived the comradeship of the trenches, not only for those who had served but also for members of the next generation, who had been too young to fight but had vicariously experienced the war as adolescents. Disabled veterans were loudly proclaimed to be "the first citizens" of Nazi Germany, and even though their practical situation did not much improve, they were honored in monuments and at rallies. In the new Berlin that Hitler and Albert Speer intended to construct, there was to be a colossal triumphal arch, at once a memorial to the First World War (it was designed to carry the name of every fallen German soldier) and a celebration of Germany's victory in the Second. To quote Canetti again: "There is nothing that more concisely sums up Hitler's essence. Defeat in World War I was not to be acknowledged but transformed into victory . . . They [the fallen] were his masses . . . He sensed that they were the ones who had helped him to power; without the dead of World War I, he would have never existed."

Even Germans who opposed Nazism shared Hitler's hatred of the Versailles Treaty and the national humiliation it represented. Most people agreed that reparations were ruinous and unfair, that the territorial settlement in the east violated Germans' rights of self-determination, and that the limitations on Germany's military power was a hypocritical abridgment of its sovereign right of self-defense. After 1933, a broad consensus supported Hitler's efforts to return Germany to its proper place in the European society of states. But for Hitler and his most fanatical followers, revising the peace was just the first step — the means to a revolutionary new international order, certainly for Europe, perhaps for the world. This new order would not be a society of states but a collection of races competing for the space they needed to flourish. Hitler believed that the German race—the *Volk*—was destined to dominate the lands to the east, which it would rule as Europeans ruled their overseas empires—colonizing the land, exploiting its resources, and subordinating its racially inferior inhabitants.

Hitler was confident that he could overturn the peace of Versailles diplomatically, by exploiting the weakness of and divisions among the other European states, but the final goals of his racial imperialism could not be achieved without war. War was necessary to conquer the extended spaces Germany needed in the east and also to create the context for a racial revolution within Germany itself. In the ultimate struggle for the future, therefore, foreign and domestic politics would be fused, the one an essential instrument of the other, both requiring the use of extreme violence against enemies at home and abroad.

The British poet W. H. Auden called the 1930s "a low dishonest decade." It was also an extraordinarily destructive one. In addition to the violence that attended the establishment of the Nazi regime, there was unrest in France, a brief but bitter civil war in Austria, and a much more brutal conflict in Spain, where half a million people perished between 1936 and 1939. In October 1935 Mussolini launched a bloody seven-month campaign to conquer Ethiopia, in which the outnumbered and poorly equipped Ethiopian forces lost between 55,000 and 70,000 men—the Italians, 9,000. And outstripping all these in scale and viciousness was the war fought by the Soviet regime against its own population. Two million peasants and their families died resisting collectivization; some were killed, others starved, others perished on their way to exile. The "terror famine" that afflicted the Volga Valley and Ukraine in 1932 and 1933 killed somewhere between five and eleven million. No sooner had the peasant opposition been crushed than the regime turned against itself. Following the murder of the head of the Leningrad Communist Party, Sergei Kirov, in February 1934, a steadily widening net of arrests eventually spread to include a million party members and as many as seven million others; of these, a million were executed, and the rest were sent to remote forced-labor camps in what the Soviet writer and former inmate Aleksandr Solzhenitsyn called "the gulag archipelago."

The governments and the populations of the liberal democracies watched this rising tide of suffering and death with anxiety and dismay. Partisans of communism or fascism tried to explain or justify Stalin's murderous policies in the Soviet Union, Mussolini's imperialism, or Franco's brutality. A few thousand volunteers went to Spain to fight for the republic. But the overwhelming inclination among Europeans was not to get involved, to allow events to run their course with only a few formulaic protests. None of these events seemed to touch the vital interests of the other European states. None was worth risking another catastrophe like the Great War.

The violence that engulfed Europe from the Iberian Peninsula to the Russian steppes during the 1930s created exactly the right atmosphere for the political offensive that Hitler was planning in Berlin. On the evening of February 3, 1933, just four days after becoming chancellor, he met privately with the leaders of the German army, who were pleased by his commitment to expanding the military, although perhaps a bit unsettled by vague references to the conquest of "living space" in the east. In October, Hitler withdrew the German delegates from the international disarmament conference, a frail survivor of the optimistic 1920s, and from the League of Nations. Two years later, he announced that Germany would no longer abide by the restrictions imposed on its armed forces by the Versailles Treaty. And then, in March 1936, he sent troops into the Rhineland, where a demilitarized zone had been established in 1919 to protect France from a surprise German invasion. The postwar settlement, already successfully challenged by Japan in Manchuria and by Italy in the horn of Africa, was now collapsing in the heart of Europe.

The forces of order proved impotent against these various assaults on the status quo. Britain and France, as members of the League of Nations and as the guarantors of the Versailles settlement, protested but did not act. What other response was possible when no one, in government or in the public at large, was prepared to risk another world war? "Anything rather than war!" the French novelist Roger Martin du Gard wrote to a friend in September 1936. "Anything! . . . even Fascism in Spain . . . even Fascism in France: Nothing, no trial, no servitude can be compared to war: Anything, Hitler rather than war." A year later, the great philosopher Bertrand Russell told a cheering crowd in London that if the Nazis were foolish enough to invade Britain, they should be welcomed like tourists, since even successful resistance would cause more damage than peaceful conquest. "The Nazis would find some interest in our way of living," Russell added, "and the starch would be taken out of them." While few Europeans were as consistent as Martin du Gard and Russell, most were firmly convinced that none of the issues facing them—the fate of Manchuria, the creation of a new German army, the conquest of Ethiopia, the remilitarization of the Rhineland—was worth a war that would again infest their private lives and perhaps bring European civilization to an end.

On November 5, 1937, Hitler summoned his foreign minister and the leaders of the army, navy, and air force to the newly completed Chancellery for a discussion of his long-range objectives. The four-hour meeting, of which we have a summary prepared by Colonel Friedrich Hossbach, the Führer's adjutant, was not, as historians have sometimes claimed, a road map for war. But it did clearly formulate Hitler's central goal, which was "to make secure and to preserve the racial community and to enlarge it." This, he insisted, was a question of space. Three of Hitler's listeners—the war minister, Field Marshal von Blomberg, the commander of the army, General von Fritsch, and the foreign minister, Baron von Neurath—expressed some misgivings about these ambitions. Within a few months, they had all been replaced by more pliable subordinates.

On November 9, Neville Chamberlain, prime minister of Great Britain since May, discussed the international situation in London's Guildhall. Both the speaker, a member of a politically prominent family with deep roots in Britain's commercial elite, and the occasion, an annual gathering of the City's business leaders, belonged to a world vastly different from the meeting in Hitler's grandiose Chancellery. This was Norman Angell's world, a world of calculation and bargaining, contracts and compromise, production and consumption. Here the choice between peace and the "perpetual nightmare" of war was obvious: "One has only to state these two alternatives," Chamberlain told his audience, "to be sure that human nature, which is the same all the world over, must reject the nightmare with all its might and cling to the only prospect which can give happiness." Three days later, a committee of the British military command provided another compelling reason to avoid war: despite the assistance Britain might get from potential allies, the committee's report concluded, "we cannot foresee the time when our defense forces will be strong enough to guard our territory, trade, and vital interests against Germany, Italy, and Japan simultaneously."

In March 1938, Hitler moved beyond the borders of the Reich for the first time. Taking advantage of an active Nazi movement in Austria, he forced the Austrian government to accept an incursion of German troops and the merger of the two countries. The so-called *Anschluss,* which Hitler accomplished with apparent ease and to the cheers of many Austrians themselves, was greeted with only mild public objections by the other powers. Alexander Cadogan, an undersecretary in the British Foreign Office, wrote to his ambassador in Berlin, "Thank goodness Austria's out of the way . . . I can't help thinking that we were badly informed about opinion in that country . . . I can't work up much more indignation until Hitler interferes with other nationalities."

This, of course, was exactly what Hitler intended to do — as Cadogan was probably aware. His next step was Czechoslovakia, the last remaining parliamentary regime in eastern Europe, which had a large and increasingly pro-Nazi German minority. Although Czechoslovakia looked formidable enough, being allied with both France and Russia, Hitler hoped to use this minority to break up and eventually destroy the Czech state. In May 1938, when the Czechs responded to German threats by mobilizing their army and securing the apparent support of Britain and France, Hitler backed down. But he did not abandon his goal. "It is," he told his army commanders, "my unalterable decision to smash Czechoslovakia by military action in the near future."

That summer, ethnic violence in the Czech lands increased, encouraged by a strident propaganda campaign orchestrated from Berlin. War seemed unavoidable. In a desperate effort to avert catastrophe, Prime Minister Chamberlain made three trips to see Hitler in September: the first, to his mountain retreat at Berchtesgaden, was apparently successful in finding a compromise solution; the second, to the Rhenish town of Bad Godesberg, ended with Hitler's withdrawing his previous concessions; and the third, to Munich, essentially gave the Führer everything he wanted. When Jan Masaryk, the Czech ambassador to Britain, learned what Chamberlain intended, he replied, "If you have sacrificed my nation to preserve the peace of the world, I will be the first to applaud you, but if not, gentlemen, God help your souls."

Chamberlain's policy of "appeasement" has become synonymous with weakness, as "Munich" has long been a symbol for the inevitable catastrophes that result when aggression is not resisted. Since 1945, the "lessons of Munich" have been cited again and again to demonstrate the need for a resolute response to a variety of foreign political antagonists. Whatever the merits of Chamberlain's policy in 1938, it is well to remember that, in itself, appeasement is not necessarily a bad thing. In their efforts to resolve conflicts without resorting to violence, diplomats often must appease their opponents. It is also important to recognize that the alternative to appeasing Hitler in 1938 was fighting him. He was not bluffing, and the *threat* of war alone would not have stopped him, as is shown by the brevity of his tactical retreat in May.

The question then becomes, Would it have been better to fight in 1938 than in 1939? None of the powers, including Germany, was prepared for war in 1938. While the Allies would not have had an easy time defeating the Germans that fall, it is very likely they would have had a better chance then rather than in 1939 or 1940. But how to defeat the Germans was not the question on Chamberlain's mind when he flew to see Hitler; he wanted to avoid fighting at almost any cost. And that was the source of appeasement's failure as a policy. It was not wrong to seek a peaceful solution to the Czech crisis, but it was criminally irresponsible not to consider what to do if that didn't satisfy Hitler's appetite for conquest.

One thing about appeasement in 1938 is abundantly clear: the overwhelming majority of Europeans were delighted when the policy seemed to work. A few skeptics warned of the disasters that lay ahead, but their voices were drowned out by the cheers of those who felt what the French socialist Léon Blum called "cowardly relief" that they had been spared another war. When Édouard Daladier, the French premier, flew back to Paris from Munich, he expected the crowd at the airport to be hostile. Instead, they had come to applaud him. Even a year after Munich, in the first public opinion poll ever taken in France, 57 percent of the population still viewed the agreements favorably. Chamberlain was given a hero's welcome when he returned to Britain with the promise of "peace in our time," and, much to Hitler's dismay, the prime minister was also popular in Germany, where most of the citizenry was relieved to see the danger of war recede.

The Munich agreements did not last six months. In March 1939, German troops entered Prague, completing the destruction of the Czechoslovak state. For the first time, Hitler could not invoke the principle of self-determination to justify his assault on the international order. This was unabashed aggression, without the pretext of an embattled German minority to protect. Even before the invasion of Prague, public sentiment in France and Britain had begun to move away from appeasement, although their governments clung to the hope that peace might still be possible.

For his part, Hitler was set on war. No sooner were German troops in Prague than he gave orders to prepare for an invasion of Poland, calling for a level of destruction that would eradicate the Polish state from the political landscape. At this point war was inevitable; the Poles, unlike the Czechs, would fight. In a last desperate effort to deter Hitler, Chamberlain signed an alliance with the Polish government, promising assistance in case of attack. France was already committed to come to Poland's aid. Hitler, however, had no reason to think that this time the western powers were serious, especially after he succeeded in signing an alliance with the Soviet Union, which deprived the Allies of a potential partner and ensured that the Poles would confront enemies on both east and west.

Even as German forces were crossing the Polish border, some statesmen in Britain and France would have been willing to make a deal. On September 2, after hostilities had begun, the French government made one last, fruitless attempt to find a peaceful way out, but by then there were no alternatives to war. Chamberlain's speech to the House of Commons on Sunday, September 3, was a grim expression of regret, not an inspiring call to battle: "This is a sad day for all of us, and to none is it sadder than to me. Everything that I have worked for, everything that I have hoped for, everything that I believed in during my public life, has crashed into ruins." Dwight Eisenhower, a forty-nine-year-old colonel in a dead-end job on

Douglas MacArthur's staff in the Philippines, heard Chamberlain's speech on the radio in Manila. "It is a sad day for Europe and the whole civilized world," he wrote in his diary. "It doesn't seem possible that people that proudly refer to themselves as intelligent could let the situation come about."

Everywhere in Europe, the popular mood in 1939 was dramatically different from what it had been in 1914. Writing from Paris, William Bullitt, the American ambassador, noted that "the whole mobilization was carried out in absolute quiet. The men left in silence. There were no bands, no songs, no shouts of 'on to Berlin' or 'Down with Hitler.'" The prefect of the Rhone described the public mood in his region as "something between resolution and resignation." Much the same was true in London, where people expected German bombers at any moment and had begun to evacuate their children to the countryside. The dean of Durham Cathedral, recalling the nationalistic fervor at the start of the previous war, noted that in 1939 patriotic sermons had to be toned down so as not to offend the congregation.

Even in Berlin, where the dogs of war had been unleashed, the atmosphere was subdued. When Hitler went to the Reichstag to announce the start of the war, he drove through empty streets. There were no cheering crowds, no spontaneous expressions of joy and solidarity, no smiling girls throwing flowers. However much they might hope Hitler would bring them victory, Germans, like the rest of Europe, knew—or thought they knew—what it might cost.

W. H. Auden was in New York on September 1, 1939, at once depressed and relieved to be so far away from the war. In the poem he wrote to describe his feelings on that day, he describes what lay ahead:

> The enlightenment driven away,
> The habit-forming pain,
> Mismanagement and grief:
> We must suffer them all again.

The saddest word is the last.

6

* *

The Last European War

T HE LAST EUROPEAN WAR began on September 1, 1939, when the German battleship *Schleswig-Holstein* bombarded the Polish garrison on the Westerplatte, a peninsula in Danzig's harbor. At the same time, three German army groups smashed into Polish territory and, in little more than a week, drove the outmatched Polish army back to the outskirts of Warsaw. The campaign's strategic objective, Hitler instructed his military commanders, was "not the arrival at a certain line," but rather "the elimination of living forces"—in other words, the destruction of Poland as a state, society, and culture. The coming war, he insisted, was to be fought "with the greatest brutality and without mercy." As they swept across Poland, German troops responded with unrestrained ferocity to anyone or anything in their path: prisoners were shot, villages burned, hostages seized and executed. We can see in the first weeks of the war the extraordinary violence that would engulf much of the continent for the next six years: the wanton killing of civilians, the targeted murder of political leaders, and the initial steps in the Nazis' war against the Jews. Much of this violence was carried out by special units of the SS and Gestapo, but the regular army was also involved, revealing a capacity for ruthlessness and brutality that would increase as the war went on.

There is no better place to observe the war's painful ironies and moral perplexities than Poland. Britain and France had declared war on Germany in order to defend their Polish allies against Hitler's aggression, but then did nothing to help them militarily. After making a brief incursion into German territory in mid-September, the French army withdrew to its fortified positions; in the first four months of the war, three British soldiers were killed on the western front. Instead of getting assistance from the west, the Poles were attacked from the east when the Soviet Union, acting on the basis of the secret annex to its Non-Aggression Pact with Hitler, occupied a large portion of Polish territory. Against these two invaders the Polish army fought bravely but in vain; by early October, military operations were over, leaving a hundred thousand Polish dead and a million prisoners. Following the defeat of their regular army, millions of Poles continued to resist the German occupation, both in an underground Home Army and as members of the Allied forces. At the end of the war, despite the Poles' courageous and consistent opposition to Nazism, the western Allies entrusted their future to the Soviet Union, under whose auspices a new set of tyrants arrived, this time wearing the ill-fitting costume of liberators. "The fate of Poland seems to be an unending tragedy," Winston Churchill told the House of Commons in June 1946, "and we, who went to war on her behalf . . . watch with sorrow the strange outcome of our endeavors."

"The very situations that bring about a modern war," the French political philosopher Raymond Aron observed, "are destroyed in its wake." Europe went to war over Serbia in 1914 and over Poland in 1939, but who among their original champions cared about either of them by the time the war was over? More important than why the war began or even how it ended was the way it was fought, what Aron called "the battle in and for itself."

Three aspects of the way the Second World War was fought were especially crucial for Europe's future. First, in the Second World War, even more than in the Great War, the nature of the battle was determined by one participant, Germany. Without Germany, there would have been no European war in 1939. Germany's amazing military performance—in achieving early victories and postponing eventual defeat—shaped the war's course and outcome. It is not surprising, therefore, that the "German question" dominated European politics in the years right after the war and remained significant throughout the rest of the century. Second, the war was fought very differently in the east and west. From the start, the Germans were determined not simply to conquer their eastern enemies but to destroy their social and political institutions so that they could impose a form of colonial domination. For Germany's enemies and allies both, the war in the east was fatal: between the Soviet frontier and the Rhine, no prewar regime remained in 1945. Third, the Second World War was a total war in which the line between combatants and noncombatants, already violated in the first war, was erased, not only during the land war in the east but also by the strategic bombing of civilian targets. The result was a war of unparalleled destruction that transformed Europe's physical and cultural landscape. It is difficult to find in all of European history a conflict with greater consequences than the one that began in Danzig's harbor in 1939 and ended, six years later, in the ruins of Berlin.

Once Poland had been defeated, Hitler was eager to get on with the larger project of racial imperialism, which meant moving into the vast territory of his temporary ally, the Soviet Union. First he had to secure his western flank by defeating France and driving the British from the continent. On October 10, as soon as the fighting in Poland stopped, Hitler ordered his military commanders to prepare a western offensive that would begin as soon as possible. A combination of factors—the generals' reluctance to rush into a new campaign, poor weather, and the army's need to regroup and resupply—forced a postponement of operations until the following spring. While the Germans energetically applied the lessons of the Polish campaign, the British and French armies did nothing, lulled into complacent passivity by eight months of the so-called Phony War in the west.

The western Allies waited behind the Maginot Line, an elaborate complex of fortifications that ran for 140 kilometers along the Franco-German border. Had it been combined with a creative strategy, the Maginot Line might have contributed to French security; instead, it became an excuse for lassitude and a screen for incompetence. As they settled into their billets, tended their gardens, and waited for the Germans to take the initiative, Allied troops lost both mobility and motivation. When German troops began to move west in the early morning hours of May 10, the French High Command miscalculated the location of the offensive's main thrust, sent its best troops north into Belgium, and allowed the enemy to drive through the Ardennes Forest, which the Maginot had left unprotected. In the confusion that followed, the French were unable to counterattack the long, exposed flank of the German armored columns as they moved swiftly toward the English Channel. Within two weeks, the French army began to disintegrate. The sour smell of defeat was in the air. An armistice, signed in the same railroad car that had been used in November 1918, was unconditionally accepted by France on June 22. The same army that had once stood firm against Germany for four long years collapsed in as many weeks twenty years later.

How could this have happened? The most obvious explanation was military incompetence, particularly in the High Command. The leaders of the French army were, in the French historian Marc Bloch's words, "incapable of thinking in terms of a new war ... The German triumph was, essentially, a triumph of intellect." Had French troops been led with more skill and imagination, they could have given a much better account of themselves and even, with a little bit of luck, been able to disrupt the Germans' risky blitzkrieg strategy. But Bloch, like many of his contemporaries, concluded that the speed and scale of the French defeat, attended by governmental paralysis and chaotic scenes of fleeing refugees, reflected a deeper social and cultural malaise. As the poet Paul Valéry wrote in June 1940, "France expiates the crime of being what she is," by which he meant soft, self-indulgent, and divided. Marshal Pétain, the hero of Verdun who became the leader of the collaborationist Vichy regime, bitterly described France as a society in which "the spirit of enjoyment" had triumphed over "the spirit of sacrifice."

Following the catastrophe of the first war, French men and women had good reason to think that they had sacrificed enough — and even that they had earned a little joy. Like every other democracy, France had neglected its security needs after 1919, in part because ordinary people, as well as their leaders, could not bring themselves to believe that anyone would be foolish enough to inflict another major war on Europe. Memories of the terrible costs of that war had surely helped to encourage the overwhelming desire for peace at any price that pervaded French society after the collapse of the army in 1940.

The debacle of 1940 was not just a French defeat. It included all the democratic states of western Europe — the Belgians and the

Dutch, who had fecklessly tried to preserve their neutrality before 1939 and were then swiftly overrun by the Wehrmacht, and the British, whose army was no more successful than the French in responding to the German assault. In contrast to 1914, when the relatively small British Expeditionary Force had played a vital role in the stabilization of the western front, in 1940 British troops were pushed back to the Channel ports, from which the best they could do was to escape to Britain, leaving their weapons and heavy equipment behind. That the precipitous withdrawal from Dunkirk was regarded as a sort of victory suggests just how desperate Germany's opponents were for good news during that perilous summer. By the end of June, only the Channel prevented Britain from suffering the consequences of its political errors and military failure.

After the fall of France, Hitler may have hoped that Britain would accept a role of subordinate neutrality that would allow him to pursue his primary objectives in the east. Winston Churchill, however, who had become prime minister on May 10, was committed to staying in the war. He was supported by the overwhelming majority of the British public, some of whom were unreasonably relieved that they no longer had to worry about allies as weak and unreliable as the French. The Germans now had to compel the British to surrender, either by striking fear into the population with airpower or, as a last resort, by carrying out a cross-Channel invasion. After three months of bitter fighting in the skies above the British Isles, the Royal Air Force managed to defeat the Luftwaffe and force Hitler to postpone further military operations.

While the air battle over Britain was still raging, Hitler began to plan an attack on the Soviet Union. Once the Soviets were out of the way, he believed, the British would have to see the folly of continued resistance, which rested on the hope that eventually Russia would come into the war on their side. He was convinced, moreover, that victory in the east would be quick and easy. If the Wehrmacht had been able to destroy the French and British armies in a few weeks, they could surely do the same to the Red Army, which had had an embarrassingly difficult time defeating tiny Finland during the winter of 1939–40. In July, Hitler ordered his military staff to prepare for a massive campaign, designed not simply to defeat the Soviets in battle but also to destroy their political system. This was not to be a war that traditional strategists would have recognized, a war in which two states fought for political advantage; this was a war of conquest and annihilation, comparable to what Europeans had inflicted on the colonized world. "Russia," Hitler is supposed to have said, "will be our India."

Stalin, who had murdered thousands of his own countrymen because he feared they might oppose him, was curiously reluctant to believe that Hitler was planning to break their agreement. Despite an increasing number of intelligence reports on Germany's military preparations, Stalin feared he was being drawn into a trap by the British and Americans, who, he was convinced, wanted to see Soviet soldiers die to save capitalism. When German units pounded across the frontier on June 22, 1941, it was a terrible shock and, by any measure, a stunning disaster. Over the next six months, the Soviets lost 8,000 aircraft, 17,000 tanks, and 4 million men. Smolensk and Kiev were occupied, Leningrad was encircled, and Moscow was besieged; many of the country's richest agricultural and industrial regions fell into German hands. On October 4, Hitler, just back from his headquarters in the east, told a jubilant crowd in Berlin's Sportpalast that "the greatest battle in the history of the world" had destroyed the Soviet Union. Moscow, defended only by a thin line of weary and hastily assembled troops, would surely capitulate. The city, he had privately decided, was to disappear from history, totally submerged under a huge artificial lake.

Despite its initial success, the German invasion did not destroy the Soviet system or the Russian people's will to resist; the national pulse continued, faintly and erratically, to beat. A renewal of the German offensive in October bogged down west of Moscow, where Stalin had decided to remain. When winter began, it was apparent that the Germans had been unable to overcome two closely related problems: one was the logistical problem of a constantly extending supply line, which frequently left troops without fuel, ammunition, and other critical supplies; the other was the operational problem of coordinating armor and infantry, which gave some Russian units the opportunity to escape. Contrary to the conventional image of blitzkrieg, only a few German divisions were motorized and the rest moved on foot; in June 1941, the invaders had 3,350 tanks, but 650,000 horses. In a theater of operations as small as France, these problems were manageable; in the east, they inhibited the German advance just enough to open a slender escape route through which the Soviets were able to squeeze.

The turning point of the war came in the first week of December 1941, when two events, occurring thousands of miles apart, tipped the balance of power against Germany. The first took place in the freezing dawn of December 5, deep inside the Soviet Union, when Russian troops launched a counteroffensive against the Wehrmacht. This did not, as Stalin hoped, rout the invaders, but it did break the siege of Moscow and reveal the Red Army's astonishing recuperative powers. The second transformative event happened two days later, on a bright Sunday morning in the Pacific, when Japanese fighter-bombers attacked the American naval base at Pearl Harbor. On December 11, Hitler joined his Japanese ally by declaring war on the United States. In his memoirs, Churchill recalled that when he learned of the attack on Pearl Harbor, he thought, "So we had won after all . . . There was no more doubt about the end . . . Being saturated and satiated with emotion and sensation, I went to bed and slept the sleep of the saved and thankful."

The survival of the Soviet Union and the engagement of the United States had, for the first time since 1939, made Germany's defeat a real possibility. For this to happen, the Soviets would have to stay in the war, and the sea lanes linking the United States and its European allies would have to remain open. Throughout 1942 and into 1943, therefore, the outcome of the war depended on the battles raging in Russia and the Atlantic. A loss on either front would have put the Allies' entire military enterprise in jeopardy.

"The only thing that ever really frightened me during the war," Churchill wrote in his memoirs, "was the U-boat peril." Churchill had every reason to be frightened. For three years, Germany seemed to be winning the battle of the Atlantic. Between June and September 1940, U-boats sank 274 ships, losing just 2 of their own. In 1942, with the entrance of the United States into the war and a significant expansion of Atlantic traffic, Allied losses mounted at an alarming rate, in large measure due to the American navy's refusal to take even the most obvious precautions, such as imposing blackouts along the eastern seaboard, where the bright lights of coastal cities made ships into perfectly silhouetted targets for the waiting packs of submarines. At first, even the Allies' ability to read Germany's coded messages was not enough to overcome their enemy's skill and their own errors, but gradually the situation improved. By February 1943, more ships were being produced than lost; in May, enough U-boats were being sunk that the German High Command was forced to limit their activity. The fighting in the Atlantic was a classic war of attrition, in which slow, incremental gains gradually raised the cost beyond what the enemy was willing and able to pay.

In June 1942, while the sea battle was still going badly for the Allies, President Franklin Roosevelt told Secretary of the Treasury Henry Morgenthau, "The whole question of whether we win or lose the war depends on the Russians." At this point, the odds of Russia's survival were no better than even. In May, Stalin had ordered an ill-prepared and badly executed offensive that produced 100,000 Soviet fatalities and another 200,000 prisoners. The Germans responded with an offensive of their own, which finally halted when General Friedrich Paulus's 6th Army failed to take the city of Stalingrad. Here both sides engaged in incredibly brutal, often hand-to-hand combat. In February 1943, after the besieging German forces were themselves encircled, the battered remnants of the 6th Army were forced to surrender, thereby inflicting a serious and symbolically charged defeat on the German campaign. Even more devastating than Stalingrad was the huge tank battle fought at Kursk in July 1943, which resulted in losses of men and materiel from which the Germans never recovered.

Kursk was just one of several disasters to befall Germany that month. At about the same time, the British and Americans, after having driven the Italians and Germans from North Africa, successfully landed on Sicily. At the end of the month, a coup removed Mussolini from power. July 1943 also saw one of the worst air raids of the war, the firestorm in Hamburg that killed thirty thousand civilians and left half a million homeless. The only war Hitler might have won—the war in which better preparation, strategic surprise, and superior operational performance gave the Germans a decided advantage over their opponents—was over. Germany was now forced to wage a different kind of war, facing an array of enemies with much larger populations, many more resources, and far greater productive capacities. In 1943, for example, Germany produced

24,807 aircraft, 270 ships (all submarines), 17,300 tanks, and 27,000 artillery pieces. The combined figures for the United States, Britain, and the Soviet Union were 147,161 aircraft, 2,891 ships, 61,062 tanks, and 210,044 artillery pieces.

After the summer of 1943, the most interesting question was not Why did the Allies win? but rather Why did it take the Germans so long to lose? Consider the course of the war over the next two years: following its defeats in Russia, North Africa, and southern Italy, the German army was gradually pushed back on every front; the German submarine fleet was being sunk; and the air force was destroyed in a hopeless attempt to defend German cities from Allied bombers. In the three months following the successful Allied landings in Normandy, in June 1944, more than a million German soldiers were killed or captured. An Anglo-American army occupied Rome and another was racing across France. The Russians were on the banks of the Vistula and had seized much of southeastern Europe. Yet the Germans continued to inflict heavy casualties on their opponents, forcing them to pay dearly for each victory. In Sicily, for example, a German force numbering about sixty thousand, operating without air or naval support, held off an army eight times larger for thirty-eight days and then managed an orderly withdrawal with most of its troops and equipment.

Why did the Germans keep fighting so long after any chance of victory was extinguished? Some may have believed that the Führer would save them, perhaps with a new secret weapon—like the jet fighters and rockets that came into operation in the last months of the war—which could turn the tide in their favor. Hitler and those closest to him clung to the hope that the enemy alliance would shatter. But even for those without such faint hopes, there seemed no choice but to go on fighting. The Nazi leaders, and a great many other Germans besides, knew what they had done—knew about the mass murder of Europe's Jews, the wicked occupation policy in the east, and the millions of Soviet POWs who had died from mistreatment and neglect. Many realized that they would not outlive the regime, which was now the only thing standing between them and the rage of those who had suffered at their hands. Since the Nazi leadership could not surrender to the enemy, there was nothing to do but surrender to the war itself.

But what kept ordinary German soldiers in the field, sometimes outnumbered seven to one, usually without air support, often with limited supplies of essential equipment? For some, ideological commitment was surely important. Those in the elite units of the SS, as well as many young people who knew no other world but the Third Reich, fought on because of their unwavering belief in Hitler, the destiny of the race, and the evil of its enemies. Others fought for the same reason soldiers always do, for their comrades. One of the reasons for the German army's effectiveness was its ability to create the small-unit cohesion that, as the American military thinker Edward Luttwak has written, "survives the terrible impact of battle far better than any other individual source of morale." And, not least important, men fought because they were afraid not to, afraid not only of being captured by the Russians, but also of being executed for desertion. Unlike the First World War, when the German army killed a small number of its soldiers, in the second, the death penalty was meted out with increasing frequency. About 15,000 men were executed on the eastern front—equivalent to an infantry division—and in the west, deserters were still being strung up from lampposts when Allied forces were a few streets away.

The Second World War, like the first, was a war of attrition, which was ultimately decided by the slow, difficult progress of the winners and the gradual, painful decline of the losers. The critical difference between the two world wars was not their character, but their scale. The second, much more than the first, was a truly global conflict, fought across the Eurasian land mass, on all of the earth's great oceans, and in the air above the entire continent. Individual battles engaged enormous numbers of men and machines: at Kursk, for instance, 7,000 tanks fought one another in an area half the size of England; in preparation for their final offensive, the Soviets amassed an army of 6.7 million men along a front stretching from the Baltic to the Adriatic.

Nothing like this had ever existed in the history of warfare. However voracious the first war's appetite for blood and treasure, the second consumed much more—more lives, more resources, more machines. Some fifty million people throughout the world died in the war. Huge quantities of ammunition were expended—the Soviets fired more than a million shells on the opening day of their assault on the Seelow Heights in April 1945. Machines were destroyed at an extraordinary rate: a Soviet field gun might last for eighteen weeks, tanks and aircraft for about three months. During the heaviest fighting in 1941 and 1942, the Soviets lost half their aircraft and one tenth of their armor *every week.*

Far more than the first, the second war engulfed soldier and civilian alike. Between 1914 and 1918, the war's destructive power was largely concentrated on the battlefield; the overwhelming majority of dead and wounded were soldiers. Between 1939 and 1945, battle had no boundaries; both sides intentionally targeted their opponent's civilians, who died in greater numbers than those in uniform. During the siege of Leningrad alone, 650,000 men, women, and children perished, more than the combined military fatalities of both France and Britain in the entire war. At war's end, most of Europe's great cities had been damaged, some — Berlin, Dresden, Warsaw, Budapest, Belgrade — had been virtually destroyed.

In wars of attrition, no single battle is decisive, but everything counts that can eliminate some of the enemy's assets and undermine his ability to fight; there is no strategic distinction between the people who manufacture tanks and those who drive them into battle. Every wartime leader, therefore, shared the sentiment of the American officer who declared, "There are no civilians in Japan . . . We intend to seek out and destroy the enemy wherever he or she is, in the greatest possible numbers, in the shortest possible time."

Aerial bombardment provided the most effective means of destroying the enemy, wherever they might be. In the 1920s and 1930s, the strategic use of airpower had appealed to military thinkers because it promised to break the catastrophic stalemate of the First World War. In works such as *Command of the Air,* published by the Italian airman Giulio Douhet in 1921, advocates of aerial bombardment argued that by restoring speed and mobility to war, the airplane could make victory possible once again. Like the advocates of the offensive before 1914, these strategists emphasized the psychological component of war — not, of course, the morale of advancing troops, but rather the morale of a civilian population which would break when death and destruction rained down from the sky. Contemporary society was thought to be particularly vulnerable to such attacks, because, in Liddell Hart's words, it was "such a complex and interdependent fabric." Liddell Hart was convinced that a "sudden and overwhelming blow from the air" could paralyze "a modern civilized state" in a few hours, at most in a couple of days.

The same aspects of airpower that so appealed to its advocates terrified many Europeans. In a frequently quoted speech to the House of Commons in 1932, Prime Minister Stanley Baldwin declared that "the man in the street" had to realize that nothing could protect him from being bombed. "Whatever people may tell him, the bomber will always get through." The only defense, Baldwin concluded, was to strike sooner and harder than your opponent, which simply means that "you will have to kill more civilians, more women and children first, if you want to save yours from the enemy." No wonder that the airplane reminded people of another famous invention that threatened to destroy its creator: "Mankind is Frankenstein," wrote the British peace activist Helena Swanwick in 1934, and "science, especially the science of aviation, is his monster. Can we learn to control it?" Throughout the 1930s, and especially after 1935, when Nazi Germany began to build an air force in violation of the Treaty of Versailles, the specter of burning cities and broken societies made the search for peace all the more urgent.

The British, whom the Channel had shielded from the direct experience of war, were particularly fearful of the bomber. By the time the war began in September 1939, they had already evacuated 1.75 million women and children from vulnerable urban areas; when the air raid sirens went off on September 3, not long after Chamberlain's somber announcement of the start of hostilities, people expected the worst. In fact, Britain and France initially refrained from attacking civilian targets, and even the Germans, who had mercilessly bombed Warsaw, restrained themselves in the west.

This could not last. When Hitler began to bomb Britain in the summer of 1940, the British responded. Without troops on the continent, they had no other way of striking at Germany. Even if the antagonists had wanted to concentrate on military targets—which they did not—it would have been technologically impossible to do so. In the summer of 1941, British investigators found that only about one third of their side's bombs fell within five miles of their intended targets, while others were as many as seventy-five miles off.

By 1942, the British had abandoned all pretense of precision bombing and were intentionally hitting cities in order to disrupt civilian
society. The Americans clung to the illusion of accuracy somewhat
longer, but they, too, eventually accepted the necessity of striking
large areas rather than specific targets. In place of precision, Anglo-
American bombing strategy substituted magnitude, steadily increasing the size of the bombs and the number of bombers. This made
it possible to conduct giant raids, such as the firebombing that flat-
tened five square miles of Hamburg in July 1943.

Airpower did not fulfill the aspirations of its most extravagant
advocates. Until very late in the war, the German economy continued to function; in fact, the production of war materiel increased between 1943 and 1944 as the bombing intensified. And while bombing caused a great deal of suffering and dislocation in the civilian
population, it did not, as Douhet and others had predicted, destroy
morale or encourage resistance. By making people dependent on the
authorities for the necessities of life, bombing may even have made
them less willing to act independently. Nor was it true that there
was no defense against aerial attack. Some bombers always did get
through, but many, on both sides, were shot down, either by enemy
fighters or by antiaircraft guns. Being a member of an Allied bomber
crew was one of the war's most dangerous assignments.

Nevertheless, after the cost of the air war has been calculated and
its strategic limitations taken into account, there can be no doubt
that it contributed substantially to the Allied victory over Germany
and, even more decisively, over Japan. In specific instances, such
as its role in delaying the development of Germany's rocket program, bombing had a direct and manifest impact. More generally,
it drained resources that the Germans would have used elsewhere;
German war production increased in 1944, but without the bombing
it would have increased still more. Finally, by attacking targets that
the Germans had to defend, the bombing campaign against cities
helped to degrade and eventually to destroy the German air force.
If the Wehrmacht had had air support in its long, bloody defense of
the Reich between the Normandy landings and the end of the war,
the Allies' victory would have taken longer and cost many more lives
on both sides.

While hundreds of British and American airmen were fighting

their war of attrition in the sky above German cities, millions of Soviet soldiers were fighting theirs on the ground and at home. On the eastern front the difference between soldiers and civilians had as little meaning as it did in the burning buildings of Rotterdam or Coventry, Hamburg or Dresden. For three years, from June 1941 to the summer of 1944, the western regions of the Soviet Union were the setting for some of the most vicious combat in the history of warfare. The German invasion, the Soviet retreat, and then the slow process of reconquest had ghastly consequences for those unfortunate enough to get in the way. Both sides took what they needed from the civilian population and destroyed everything else of value. In the depths of winter, German troops drove peasants from their homes, set fire to their villages, put the strongest to work repairing roads or building tank traps, and left the rest to fend for themselves. Even on wartime Europe's broken and bloody terrain, the suffering of the Soviet people stands out. The figures are necessarily inexact, but close enough to suggest the magnitude of the tragedy: 6.8 million military dead, about 16.9 million (and perhaps as many as 24 million) civilian dead. Of the 50 million people who lost their lives in the war, therefore, at least half were citizens of the Soviet Union.

It is against this background of devastation that the Soviet achievement must be calibrated. The core of that achievement was military: the extraordinary ability of the Red Army to absorb the punishment inflicted by Germany and then slowly to rebuild and recover. Behind the Soviets' steadily improving military performance was an impressive feat of economic planning and production. Despite the fact that the Germans occupied much of its most productive territory, the Soviet Union was still able to produce enough to feed its citizens and supply its soldiers. Soviet engineers swiftly disassembled and moved some factories, while expanding others in areas beyond the Germans' reach. Since most men were drafted into the armed forces, the work was usually done by women, whose participation in the workforce went from 38 percent to 53 percent overall, and to over 70 percent among agricultural laborers. Soviet civilians worked more than sixty hours a week and subsisted on short rations, restricted fuel, and a bare minimum of consumer goods. With enormous effort, the Soviet economy produced simple, crude, but durable machines in very large quantities: 100,000 tanks, 130,000 aircraft, 800,000 field guns and mortars, 500 million shells, 6 million machine pistols, and 12 million rifles.

The Soviet government motivated its population by appealing to a wide range of loyalties, including traditional Russian patriotism and even Orthodox religious piety. Germany's savage occupation policies helped the Soviet war effort by alienating people who might have welcomed relief from Stalinist tyranny. The same peasants who met the invaders with bread and salt soon turned against them once they had experienced German rule. But mainly the Soviets prevailed because of who they were and the kind of system they had built—a system dominated by a rigid party structure, disciplined by an extensive terror apparatus, and energized by a totalizing set of beliefs. In wartime, these characteristics of the regime became the source of survival and success. In that sense, the entire Soviet project was a preparation for absolute war.

Reflecting on why his nation survived in 1941, a Russian general wrote: "The Third Reich's Nazi ideology and dictatorial regime achieved superiority over western bourgeois democracies; but in the east it met with a similar, perhaps even better organized regime, and fascism could not withstand the trial by fire." Both regimes were committed to a transformative vision of the future; both were ruled by a single charismatic figure and mass-based party; both enforced obedience and conformity with similar instruments of terror—denunciations, secret police, and concentration camps.

The Soviet and Nazi regimes were willing and able to demand more from their citizens and soldiers than any civilian society. The most obvious example of this was the difference in the attitude toward casualties in the Russian and German armies, on the one hand, and the Anglo-American on the other. Russian and German commanders fought more aggressively, often with wanton disregard for the human costs. If anything, this increased in the late stages of the war, when the final outcome was no longer in doubt. Soviet forces, for example, suffered 78,291 fatalities and 274,184 wounded in the battle for Berlin, many of the casualties the result of a senseless race to victory between rival commanders. German losses in the last months of the war were 1,230,045, almost a quarter of the total battlefield deaths. British and American commanders were much more frugal with the lives of their citizen soldiers. According to Chester Wilmot, one of the war's finest historians, "The gravest shortcoming of the British army" was "the reluctance of commanders at all levels to call upon their troops to press on regardless of losses." Like the British, the Americans tended to move slowly and cautiously, whenever possible employing airpower and artillery rather than manpower. Major General John Dahlquist, who wanted a more aggressive strategy, complained about the reliance on firepower that constituted the American way of war: "We cannot sit off at a distance, shell the enemy and wait for him to quit." As we know, the enemy did not quit until Soviet troops reached the center of Berlin.

The Second World War was many wars in one: a war between combatants on the battlefield, a war against enemy civilians, mainly from the air, and a tangle of subsidiary wars that raged within and alongside the larger military conflict. While these subsidiary wars were shaped by the ebb and flow of German military power, they also had deep causes and lasting consequences of their own.

Hitler had always believed that Germany was simultaneously fighting two distinct but inseparable wars, an international war against enemy states and a racial war against those alien and pathogenic agents that he believed endangered the German *Volk*. International war was a means to an end; military victory would have no purpose if it was not attended by a successful racial revolution. So between 1933 and 1939, at the same time that Hitler laid the political and strategic foundation for military victory, he created the ideological and institutional foundation for a campaign of racial domination. However imprecise the details of these plans may have been, their ultimate goal was firmly set: Germany's racial enemies would have to be destroyed and its racial inferiors conquered and enslaved.

Like the international war, the racial war was fought on several fronts at once. Among its first targets were incurably ill and socially undesirable Germans. In October 1939, Hitler signed an order, written on his private stationery, that gave certain physicians the right to grant a "mercy death" to selected patients. Similarly ruthless euthanasia programs had been advocated for decades in Germany and in many other countries, including the United States, but only the war made them practically possible. It was therefore significant that Hitler's October order was backdated to September 1, the day the war began. Initially, about 5,000 children were killed, then some 70,000 adults. At first the killing was done with injections; in late 1939 or early 1940, gas was used, sometimes in mobile vans, sometimes in gas chambers disguised as showers. Even after August 1941, when popular protests forced Hitler to order a pause in the euthanasia campaign in Germany, some scientists went on murdering patients; in the conquered eastern territories, where there was less need for caution and concealment, the campaign continued unabated.

At the core of Germany's racial war stood the Jewish question. Unlike the Poles and other Slavic inferiors, the Jews could not simply be subjugated; they were far too dangerous for that, their power too great, their influence too insidious. As early as 1919, Hitler had concluded that the only remedy was to get rid of them—although he had not yet decided how to do it. As soon as the Nazis took power in 1933, they began to strike against Germany's own Jews, who were a small, highly assimilated minority. After some initial violence, the regime acted "legally" with a series of measures that removed Jews from public office, restricted their civil and economic rights, and isolated them from the rest of society. Emigration was encouraged; by the time the war started about half of Germany's Jews had fled, leaving behind only those unable to go or unwilling to believe that their nation would ultimately reject them. As the international situation became more tense, the pace of persecution quickened. In November 1938, a few weeks after Hitler's diplomatic triumph at Munich, Nazi storm troopers burned synagogues, murdered some Jewish men and arrested many more, and broke the windows of Jewish-owned shops in what came to be called *Kristallnacht,* "night of broken glass." On January 30, 1939, in a speech celebrating the sixth anniversary of his appointment as chancellor, Hitler made the ominous prediction that "if the international Jewish financiers" succeeded in starting a war, the result would not be a triumph of Bolshevism and "thus a victory of Jewry," but rather "the annihilation of the Jewish race in Europe."

Like the euthanasia campaign and the subjugation of the Slavs, the persecution of the Jews was transformed by the war. The number of Jews under Nazi control now greatly increased, just as the war created a climate of violence and upheaval in which extreme measures, still unimaginable in peacetime, could become routine. At first the Germans uprooted, concentrated, and segregated the Jewish populations in the conquered territories, driving them from their homes into ghettos and camps. Although there was some killing, especially in the early days of the Polish campaign, until the end of 1940 only about 100,000 Jews had died—the word "only" prepares us for the magnitude of the horror to come. Following the defeat of France, the Nazi regime had still not settled on a way to rid Europe of its Jews. It seriously considered moving millions of Jews to Madagascar. The plan was abandoned when Britain, which continued to control the sea, remained in the war and the invasion of the Soviet Union brought even more Jews into the Nazi orbit. Sometime in the second half of 1941, Hitler and his closest associates decided on the so-called final solution of the Jewish question—that is, the murder of every Jew they could get their hands on.

At first the killing was done by mobile squads operating throughout the occupied east—about half of the total victims died that way —and then it shifted to concentration camps to which millions of Jews were transported from throughout Europe. Some were killed at once, others put to work until they expired. From March 1942 to February 1943, the racial war was waged with the greatest intensity; by February, between 75 and 80 percent of the Nazis' victims had died. Most of the killing was done in what had once been Poland, in camps like Treblinka, Majdanek, Belzec, and Auschwitz-Birkenau—places whose names continue to haunt our imagination. Close to six million Jews perished; another quarter million Roma and Sinti, who were added to the Nazis' extermination program in 1942, were also killed.

The scale and intensity of this murderous campaign were possible because of the resources that the Germans devoted to the racial war and the totalizing ambitions of Nazi ideology. Only a well-organized modern state, propelled by a utopian vision, could have aspired to kill every Jew in Europe, from the impoverished inhabitants of a Polish village to a family of respectable refugees hiding in an Amsterdam attic. But to succeed the Germans needed help, and almost everywhere they went, they found collaborators. Sometimes, as in Romania, the German authorities were shocked by the violence of their allies' anti-Semitism, and sometimes, as in Vichy France, they were surprised by the eagerness with which the regime anticipated their desires. The murder of the Jews was a European phenomenon, in which French militiamen, Ukrainian peasants, Latvian volunteers, and Italian Fascists all played a part.

Amid the extraordinary suffering that human beings inflicted on one another during the Second World War, the fate of the Jews is unique because of its totality. Only the Jews were to be killed to the last man, woman, and child, regardless of wealth or social position, politics or religious conviction. But while the Jews had a special status as victims, they were by no means the only targets of lethal ethnic hatred. Violent conflicts erupted between the Bretons and the French and between the Flemish and the Walloons; Poles and Ukrainians fought against the Germans and against one another, and at times they cooperated against the Jews; Czechs and Slovaks, Greeks and Albanians, Macedonians and Bulgarians, Serbs, Croats, and Bosnian Muslims all tried to establish control of disputed areas or protect themselves from attack.

Some of these conflicts were provoked by the destabilizing effect of the war, which broke down public order and empowered those with a taste for violence. But a great deal of wartime ethnic conflict was the direct result of German policy. For example, in a memorandum offering "Some Thoughts on the Treatment of the Alien Population in the East," written in May 1940, Heinrich Himmler, the chief of the Nazis' terror apparatus, argued that the occupation authorities should encourage the development of as many ethnicities as possible: "We have no interest in leading them to unity and greatness or in gradually giving them a sense of national consciousness and national culture, but rather in dissolving them into countless little splinter groups and particles." Among these ethnic groups, Himmler went on, Germany can find valuable allies who can serve as policemen and camp guards.

When Himmler began to expand the Waffen SS, the SS's military component, he formed ethnically based units both in occupied areas and from the citizens of allied states, with particular emphasis on "Germanic" races like the Dutch, Danes, Norwegians, and Flemings, but also including Latvians, Ukrainians, Croats, and Bosnian Muslims. By March 1945 only about 40 percent of the 830,000 men serving in Waffen SS divisions were German nationals. Among the defenders of the Reichstag building during the final battle for Berlin were the remnants of the Charlemagne Division, the SS unit recruited in France. By then, these desperadoes had nowhere else to go.

Yugoslavia provides a particularly vivid example of how German policy activated existing ethnic conflicts to create a complex pattern of internal warfare. Yugoslavia had been formed after the First World War when the Kingdom of Serbia joined with the Slavic provinces of the Habsburg monarchy, Croatia, Slovenia, and Bosnia. Like many of the other postwar successor states, Yugoslavia failed to integrate its various linguistic, religious, and ethnic elements. After its defeat by the Germans in a brief but savage military operation in April 1941, Yugoslavia was broken up: parts were occupied by the German and Italian armies, parts annexed or administered by Germany's allies, and an independent Croatian state was established under the rule of the Ustasha, a small Fascist movement whose leaders had found shelter in Mussolini's Italy. In an effort to "purify" Croatia, the Ustasha killed large numbers of Serbs, Jews, and Muslims.

Against the Ustasha and the Italian and German occupying forces, two competing resistance movements emerged. One, under a former Yugoslav officer, Colonel Dragoljub Mihailović, was composed largely of Serbs. The other, under a veteran Communist activist who took the name Tito, included all ethnic groups. In order to preserve his forces for the battle that he knew would begin after the European war was over, Mihailović was reluctant to engage the Germans. Because he was a much more effective and ruthless partisan fighter, Tito earned the support of the British and Americans as well as the Soviets, and, after a prolonged struggle against both Mihailović and the occupying powers, he managed to emerge victorious.

Yugoslavia was an extreme case of the tangled conflicts that occurred everywhere in Nazi Europe. All of these conflicts were multidimensional, between the Germans and their opponents, between collaborators and the resistance, and among different elements of the resistance. In all of them, the state's monopoly of violence was dissolved, the distinction between foreign and domestic disputes erased. The final outcome of these subsidiary wars depended on the conventional military campaign against Germany; nowhere, not even in Yugoslavia, was the resistance strong enough to win on its own; nowhere could the collaborators survive without German backing.

When the German armies were defeated, the regimes that had

collaborated with them were swept away. Reluctant allies like King Michael of Romania, willing partners like Father Jozef Tiso of Slovakia, enthusiastic henchmen like Ante Pavelić of Croatia, all took their well-deserved places in history's dustbin. In many parts of Europe, resistance forces exacted revenge on their enemies as soon as it was safe to do so. In France, for example, between 8,000 and 9,000 participants in the Vichy regime were hunted down and killed. Many collaborators were brought to trial and given harsh sentences —about 50,000 Frenchmen were convicted, 6,700 were condemned to death, and 767 were actually executed in the immediate postwar period. Even when the pressure for retribution eased, wartime memories remained woven in the fabric of postwar political life.

In 1945 there was no precise moment, comparable to 11 A.M. on November 11, 1918, when the Second World War ended. German representatives formally surrendered three times—on May 4 to the British on the Lüneburg Heath, on May 7 to the Americans at Reims,

The end of the Thousand-Year Reich. Nuremberg, May 1945.

and on May 9 to the Russians in Berlin—but some German units had given up much earlier, while others, like the SS troops who attacked a Soviet division in Hungary on May 20, desperately fought on for a few more days. By the end of the month, all German resistance had ceased. American fears of a Bavarian redoubt, where fanatical Nazis would gather for one last, Alamo-like stand, turned out to be groundless, as did concerns about an insurgency by "Werewolf" partisans. Considering how long and effectively Germans had fought to defend the Nazi regime, it is remarkable how swiftly and completely they quit once the regime collapsed.

Millions of German soldiers became prisoners of war in 1945; some of them perished from disease or malnourishment; the more fortunate were quickly released, and others remained captives in the Soviet Union for up to a decade. The Allies also captured several of the regime's leaders, who were placed on trial before an international tribunal in Nuremberg. Other members of the Nazi elite managed to escape, either by changing their identities and melting into the civilian population or by fleeing abroad. Perhaps the most surprising thing about the end of the Third Reich was the number of people who killed themselves rather than face the consequences of defeat. This included not only the top leaders—Hitler, Goebbels, and Himmler—but also 110 generals, many party officials, local administrators, and scores of ordinary men and women, some of whom, like Goebbels, took their wives and children with them. Without precedent in European history, these suicides are yet another expression of the intense feelings of despair, fear, and, one would like to believe, shame, generated by this most terrible of wars.

Some Germans killed themselves to escape the wrath of the invading Soviet army. While looting, rape, and murder have always followed the movement of soldiers, the behavior of Soviet troops in occupied Europe has no modern parallel. From the moment they broke out of their own territories, and especially after they entered Germany, Stalin's armies inflicted great suffering on millions of civilians. They seized everything of value and often destroyed what they could not carry away. Every German was liable to become the target of random acts of violence, but women were particularly vulnerable. Young and old, rich and poor, healthy and sick, Nazi sympathizers and resisters—hundreds of thousands were raped, often many times and over prolonged periods; thousands were killed by their assailants or died as a result of their ordeals. Rape and murder were not just perpetrated by troops fresh from the chaos of the battlefield, but continued for years after the war, throughout the Soviet occupation zone.

The war and its aftermath uprooted as many as 60 million Europeans. Millions of Jews and Poles had been driven into camps by the Nazis; the hundreds of thousands of Germans living outside the Reich had been brought "home" to be resettled in the colonized areas of the east. A half million Volga Germans, a half million Chechens and Ingush, and 200,000 Crimean Tatars were deported by Stalin "for security reasons." In the course of the war, millions became homeless when they fled from combat zones. Everywhere that armies moved, they were surrounded by crowds of civilians, burdened by whatever pitiable belongings they could carry, desperately trying to get away from the horrors that had inundated their lives. By the end of the war, close to one fourth of those living in Germany were refugees, including the 5 million people whose homes had been destroyed in the bombing ("dehoused," in the language of Britain's Bomber Command), hundreds of thousands of Germans who had fled the Red Army, as well as millions of forced laborers, prisoners of war, and the survivors of concentration camps. Some of these people returned home once the war ended, but by then the flood of refugees was swollen by 12 million Germans expelled from Poland, Czechoslovakia, Hungary, and Yugoslavia. In the immediate postwar period, these expulsions were often accompanied by acts of violence, such as the swift and brutal ethnic cleansing of Germans from the Sudetenland, in which thousands were raped, beaten, and murdered. Germans were not the only ones expelled in order to create an ethnically homogeneous territory. Hungarians were forced from lands occupied by Romania, Slavs and Albanians from Greece, and Poles from Ukraine, who were moved into houses vacated by the Germans expelled from Poland.

Some boundaries were moved after 1945, especially in what had been Polish territory between the wars, but the major changes after the Second World War involved the movement of peoples rather than borders. Eastern and central Europe's patchwork of ethnicities, formed over centuries by an endless series of migrations and new

settlements, was radically simplified. The rich and sometimes turbulent mix of religions, languages, and peoples that had once flour-
ished in cities like Salonika or regions like Bukovina was gone, the
Jews and Gypsies were dead, other minority populations had fled or
been expelled. By the end of the war, Allied leaders accepted certain
kinds of ethnic cleansing as a painful but necessary antidote to internal conflict. Churchill, for example, told the House of Commons
in December 1944 that the "expulsion [of Germans] is the method
which, so far as we have been able to see, will be the most satisfac-
tory and lasting. There will be no mixture of population to cause
endless trouble." At their meeting in Potsdam in August 1945, the
American, British, and Soviet governments recognized the need for
what they disingenuously referred to as the "orderly and humane"
transfer of the German populations out of the east.

The displaced peoples of Europe wandered through a world littered with the wreckage of war, scattered across the landscape like
some wanton child's broken toys. Along hundreds of roads could be
seen the burned-out tanks, unworkable field guns, and hastily dug
graves that marked places where men had died in some obscure skir-
mish. Europe's rivers were clogged with the debris of fallen bridges,
its harbors filled with bombed-out docks and sunken ships. Cities,
like the ancient Norman fortress of Caen, had been reduced to rub-
ble by ground warfare; from Coventry to Kiev, the urban targets of
aerial bombardment lay in ruins. To those caught in the midst of
the fighting, war has always been total, rending the fabric of civil
life and killing bystanders as well as combatants. But not since the
Thirty Years' War of the seventeenth century had such a large proportion of the European population suffered the pains of war. The
scars left on places and things and people's consciousness would remain for decades.

The war's deepest and most lasting impact was on the Soviet
Union. None of the major belligerents suffered as much human and
physical damage. None experienced a more painful transition from
war to peace, because the regime struck out against those whom it
viewed as traitors or collaborators. Chief among them were the five
and a half million Soviet citizens who were in Germany in 1945:
prisoners of war, men and women who had been forced to work in
the German economy, people who had fled westward in the hope

of escaping Stalinist tyranny, and former soldiers who had fought under the renegade Soviet general Andrei Vlasov. The secret police shot or imprisoned collaborators, but even those who had fallen into German hands involuntarily were regarded as enemies of the state. Some of them were executed; most were sent to a labor camp. Those fortunate enough eventually to return to civilian life carried a black mark on their records until 1995, when a decree by the president of the Russian Federation restored full legal rights to former prisoners of war.

The Soviets handed out the harshest penalties, but they also celebrated most vigorously their citizens' heroic achievements. No other country built as many elaborate monuments to commemorate the dead and inspire the living. In no other country was wartime service so important for establishing a political career. In the 1950s, a foreign visitor to the Soviet Union warned his colleagues, "You must resign yourselves to hearing over and over again about the experiences of your interlocutors in the Great Patriotic War." The first generation without such experiences did not come to power until forty years after the war—it turned out to be the last generation of Soviet rulers.

Britain, like the Soviet Union, had earned a place among the winners. After years of evasion and illusion, the British had stood up to the Nazis, refused to make a deal when they could, and provided the base without which a reconquest of western Europe might not have been possible. Dunkirk, the heroic young pilots of the RAF, Churchill's stirring rhetoric, the stoicism of ordinary people during the Blitz—all of these became nourishing parts of the nation's history. But unlike the Soviets, for whom the war remained a source of meaning and legitimacy, the British were eager to move on, as they emphatically demonstrated when they failed to reelect Churchill in the summer of 1945. In response to a survey conducted in 1944, most people said that instead of plaques or sculptures honoring the dead, they preferred parks and gardens, "which would be useful or give pleasure to those who outlive the war." There were, of course, parades and monuments in celebration of victory, but they were fewer in number and much less symbolically potent than the ones commemorating World War I, which the British still called the Great War.

Outside the Soviet Union and Britain, the war was an uncom-

fortable chapter in the master narrative that makes up the core of
every nation's identity. The French did their best to turn the history
of their war into a glorious story of resistance, but given the reality of defeat and collaboration, this was a myth difficult for many to
sustain. Italy, which had gone to war as Hitler's ally, made the most
of its belated movement to the winning side in 1943 and of the courageous resistance movements that had fought German troops during the war's last two years. Germans sometimes called 1945 "year
zero," a time suspended between past and present when history had
come to an end. In postwar Germany there were no memorials to
the heroic dead, no tributes to a great lost cause, no public ceremonies of commemoration or consolation. Living among the ruins,
Germans confronted the consequences of their defeat every day. Just
four years after the last shots were fired in the battle for Berlin, there
were two new German states, each claiming to represent a new beginning, a clear break from the values and institutions that had produced the catastrophe of Nazism.

Even more than the first, the century's second great war proved
difficult to commemorate. The magnitude and variety of the suffering it caused resisted canonical expression in words or stone or public ceremonies. The most moving cultural products of the war, therefore, are usually intensely individual portraits of a particular time and
place, not accounts of the war itself: Simone de Beauvoir's memoirs
of the French resistance, Beppe Fenoglio's novels about Italian partisans, Anne Frank's heartbreaking record of her life in hiding. Many
great works about the war—Heinrich Böll's novels, for example, or
Marcel Ophuls's film *The Sorrow and the Pity*—have less to do with
wartime experience than with its pervasive and elusive legacies.

One of the best collections of essays on the immediate postwar
period is called *Life After Death*. Its central theme is the sometimes
desperate attempt of people to recreate their normal lives after having been surrounded by so much destruction. For millions of Europeans the problem after 1945 was not meaning but survival—finding enough to eat, a place to sleep, a way to keep warm. To do this,
it was necessary to clear away the ruins and lay the foundations for a
new social and political order.

PART III

States
Without War

7

***

The Foundations of the Postwar World

T HE LEADERS OF what Winston Churchill called the Grand
Alliance met for the last time between July 17 and August 2,
1945, in Potsdam, an affluent suburb of Berlin. Harry Truman, president of the United States since Franklin Roosevelt's sudden death
four months earlier, had crossed the Atlantic on the cruiser *Augusta*
and traveled overland from Antwerp. Churchill, still exhausted by
the strains of wartime leadership, flew in after a brief holiday at a
borrowed château in Hendaye, a mountain town on the border between France and Spain. Ten days into the conference, when the
results of the British general elections became known, he would be
replaced by Clement Attlee, the leader of the Labour Party. Stalin,
who arrived a day later than expected, came by special train from
Moscow. Potsdam, still well within the territory occupied by the
Red Army, was as far west as he had been willing to travel.

Potsdam had been chosen for practical reasons. Although the
city had been damaged by air raids in the final months of the war,
there were enough villas to house comfortably the participants and
their staffs. Nevertheless, everyone was aware of the symbolic power
of the place, which had a long association with the Hohenzollern
dynasty, whose summer palaces were there, and with the Prussian
army, some of whose elite regiments had their headquarters in the
town. In March 1933, the Potsdam Garrison Church, where Frederick the Great was buried, had been the scene of a famous ceremony in which President von Hindenburg seemed to accept the le-

Laying the foundations. The last meeting of the
Grand Alliance, Potsdam, July 1945.

gitimacy of Hitler's claim to the legacy of Prussia's greatness. The
results of Hitler's "Thousand-Year Reich" were immediately visible
just a few miles away, in the smoldering ruins of Berlin, which still
smelled of rotting corpses. Truman returned from a quick tour of
the city deeply shaken by the scale of destruction and the magnitude
of human suffering. Stalin, who had plenty of ruined cities closer to
home, made no effort to see the wreckage of Hitler's capital.

In retrospect, the contours of the postwar era are apparent at
Potsdam. Most obvious was the deteriorating relationship between
the Russians and the Americans. The East-West conflict was still
masked by both sides' efforts to cooperate, but the difficulties of
agreement on specific issues were unmistakable. That the report
of the successful test of a nuclear device in New Mexico arrived on
the day before the conference began—Truman was given the news
just after he returned from touring the ruins of Berlin—foreshad-
owed the stakes of what would become the Cold War. The Potsdam
Conference was clearly dominated by Truman and Stalin. The British—under Churchill and then Attlee—played a peripheral part,

whereas the other European powers were unrepresented, their interests largely ignored. The foundations of postwar Europe would be laid and defended by the two superpowers, whose rivalry would divide the continent and impose upon it a new kind of order, reinforced by the awesome threat of nuclear war. Potsdam, therefore, stands at the end of one long chapter in European history and the beginning of another.

The wartime origins of conflict between the United States and the Soviet Union provide another illustration of Raymond Aron's insight that what matters most in a modern war is the way it is fought, "the battle in and for itself." When they entered the war, neither Britain nor the Soviet Union had much choice about how they would fight. They were faced with the possibility of total defeat and national extinction, so their task was the starkly simple one of survival, by any means and at all costs. In this, as in so many other ways, the United States differed from both its allies and its opponents. Protected from the combat zone by vast oceans, America did not face an immediate threat to its national existence. As a result, unlike the other belligerents, the United States had the luxury of making choices about how many troops to train, how much of its society's resources to devote to the war, and where to put the weight of its military efforts.

Even before the United States entered the war, American planners had made three important strategic decisions: in the case of a war against both Germany and Japan, the former would have priority; victory would require a substantial commitment of American ground forces; these forces would not be ready before 1943. When the plans incorporating these decisions were first drafted in 1941, they called for the mobilization of 215 army divisions, a total of 8.7 million men. By 1942, this figure had been reduced to 90 infantry divisions, which were to be supported by huge investments in air and naval power — in other words, the United States would fight a war of machines, which would minimize casualties among its citizen soldiers.

In the end, some 16 million men and women served in the American armed forces during the war, but compared to the other combatants and to its own military potential, the American army remained small — slightly larger than Japan's, smaller than Germany's, and less than half the size of the Soviet Union's. At the same time, America's productive capacities were never fully mobilized for the war effort. Even though it produced war materiel for itself and its allies, the American economy still managed to manufacture an impressive array of consumer goods. Despite shortages, rationing, and wage controls, the average American's standard of living actually improved during the war—a truly remarkable fact when one considers what was happening everywhere else.

By the time the United States entered the war, the Soviet Union had been fighting the Germans for six months. At first, American opinion about the Nazi-Soviet conflict was divided. Harry Truman, then a senator from Missouri, recommended supporting whoever seemed to be losing, so that Nazis and Communists might kill as many of the other side as possible. The Roosevelt administration, however, recognized the importance of keeping the Soviets in the war. Even before Pearl Harbor, the administration pushed a bill through Congress approving a billion dollars in aid; once the United States was in the war, the Soviets became an indispensable ally, old animosities were forgotten, public opinion shifted, and Stalin magically morphed into "Uncle Joe," an avuncular patriot rather than a bloodthirsty tyrant.

The size and pace of American mobilization meant that for three years the ground war in Europe was waged by the Soviets alone. From the Allies' strategic perspective, delaying an Anglo-American invasion of western Europe made perfect sense. But this decision had two important consequences for postwar politics. First, it made the British and Americans reluctant to press Stalin for commitments on what Europe would look like after the war. Given the constant danger that Stalin might make a separate peace with his former Nazi ally, it seemed prudent to postpone difficult issues or veil them with vague promises and elastic agreements. Second, the delayed invasion of France and then the slow progress of the Allies' drive into Germany gave the Soviets the opportunity to occupy most of eastern and central Europe during the final eight months of the war. Stalin's frequently quoted remark to the Yugoslav Communist Milovan Djilas, "Everyone imposes his own social system as far as his army can reach," was not an entirely accurate guide to Soviet policy after 1945, but it does contain an important kernel of geopolitical wisdom. It was the way the war had been fought that enabled Stalin to determine who would replace the ramshackle collaborationist regimes blown away by the winds of war. To have contested these facts on the ground would have required a military confrontation with the Soviets that even the most fervent anti-Communist was unwilling to contemplate.

From the start, Stalin had a firm notion that he wanted a Communist Europe, but how long that would take and how it might be achieved remained open. Much of the scholarship on the Cold War tends to emphasize either Stalin's long-run goals or his tactics; the former points to his ambition and aggressiveness, the latter to his flexibility and restraint. In fact, the two fit together. Both were the product of Stalin's changing perception of dangers and opportunities in the postwar international system. On a few matters Stalin was prepared to insist: Poland would have to be within the Soviet sphere; the Baltic states, which had been part of the Russian Empire before 1917, would be annexed, as would Bessarabia and Moldavia. These were all territories Stalin had acquired when he was Hitler's ally, and he had no intention of abandoning them now that he was foremost among Hitler's conquerors. Elsewhere, and especially in what had once been Britain's sphere of influence—Greece, Turkey, Iran—Stalin probed for weakness, and then retreated when he found that the United States was ready to take Britain's place. In much of the continent he pursued a variety of options: signing treaties, such as the agreement he made with de Gaulle in December 1944; supporting local Communist movements in France, Italy, and Yugoslavia; and temporarily accepting coalition governments, as in Czechoslovakia and other eastern European states.

Like the Soviet Union, the United States combined fixed goals with relatively flexible tactics. Policy makers in Washington wanted a stable, peaceful international order, free from predators like Nazi Germany and open to the spread of American values and the sale of American goods. President Roosevelt and many of his advisers hoped that this situation could be achieved in cooperation with the Soviet Union, which would be, along with Britain and China, one of the "four policemen" charged with imposing order on turbulent global neighborhoods. Roosevelt assumed that once the war had been won, the United States could, as it had in 1865 and 1919, demobilize its army and depend on ideological attraction and economic power to influence the international system. These goals expressed a broadly based consensus about the nation's real interests, as defined by its traditions, huge productive capacity, and unique geopolitical advantages. Viewed from America, the war seemed like a pathological breakdown of a system naturally based on markets and open exchange, a system that most Americans wanted to reestablish as quickly and with as little cost as possible.

The postwar division of Europe emerged gradually, not as the result of some master plan imposed by either the East or the West, but rather from a complex dynamic of conflict and adjustment in which each side, increasingly aware of its opponent's antagonistic interests and relative strength, tested the extent—and reluctantly accepted the limits—of its power.

By the time Roosevelt died in April 1945, there were already policy makers in Washington who advocated a tougher approach to Moscow. Truman, however, while not as confident about Soviet goodwill as his predecessor, continued to try to work with Stalin. Anti-Soviet voices in the administration grew louder and more insistent throughout 1946, but American policy remained fluid, subject to the diverse influences of various departments of the government. The situation changed once and for all in 1947, the year the Cold War clearly began. On March 12, Truman delivered a dramatic address to a joint session of Congress. "I believe," the president declared, "that it must be the policy of the United States to support free peoples who are resisting attempted subjugation by armed minorities or by outside pressures." Although it was an immediate response to Communist pressures on Greece and Turkey, the so-called Truman Doctrine had potentially global implications.

Three months later, Secretary of State George Marshall used a speech at the Harvard commencement to announce an ambitious if vaguely formulated plan to assist Europe's economic recovery. We now know that the Marshall Plan's biggest impact was political and moral rather than economic. Above all, it was a powerful expression of American engagement with the continent, based on the growing awareness that material want and social instability would create receptive soil for Communist subversion. The Truman Doctrine and the Marshall Plan reflected Washington's recognition that the situation in Europe and the dangers of Soviet expansion meant that the United States would have to play an active role in forming and protecting the kind of world it wanted. At the end of the year, Marshall had a tentative but still significant conversation with his counterparts in France and Britain about new security arrangements for Europe.

The year 1947 was also crucial to the evolution of Stalin's tactics, which shifted from an attempt to shape the politics of the continent as a whole to the consolidation of direct Soviet control over its eastern half. When the year began, Communist parties were still involved in coalition governments, not only in eastern Europe but also in Italy and France. One after another of these coalitions collapsed. In the West, the Communist parties went into opposition; in the East, they established one-party rule on the Soviet model, assisted when necessary by the Red Army. At the same time, a new version of the Communist International, which Stalin had dissolved during the war, was announced, as Moscow moved to impose discipline and uniformity on Communist parties everywhere. This critical phase in the division of Europe ended in February 1948 when, after a particularly blatant violation of democratic institutions, the Soviets and their local allies installed a Communist regime in Czechoslovakia.

While the western powers did as little to save Czech democracy from Stalin in 1948 as they had from Hitler a decade earlier, the Prague coup did crystallize their fears of Soviet expansion into the rest of Europe. In March, Britain, France, Belgium, the Netherlands, and Luxembourg signed a military pact in Brussels. That summer (while the Soviet blockade of Berlin was under way), the United States and Canada began confidential talks with the members of the Brussels Pact. From these discussions eventually came the North Atlantic Treaty, drafted in December 1948 and signed the following April. The treaty turned out to be a milestone in America's commitment to defend western Europe militarily, but at the time, the exact nature of this commitment was still uncertain. Consider, for instance, the studiously vague wording of Article 5, in which each signer promises to defend the security of the North Atlantic area "by taking forthwith, individually, and in concert with the other Parties, such action as it deems necessary, including the use of armed force."

After the second war, as after the first, the key to European security was Germany. But in contrast to the situation in 1918, when Germany had remained a sovereign state, in 1945 "Germany" ceased to exist, its sovereign authority devolving to the Allies, who took direct responsibility for the population living in what had been the German state. One of the few things that all of Hitler's former enemies agreed on was that Germany should never again become a threat to peace. Hitler had shown how much damage could be done when a fanatical movement took control of a well-organized state that could command the resources of a highly developed society. Many feared that amid the ruins of the Third Reich some new führer was plotting to do it again. The British and Americans, therefore, were prepared to accept the transfer of Germany's eastern provinces to Poland, which Stalin used to compensate the Poles for the territories he had annexed to the Soviet Union. Everyone agreed that Germans would have to provide reparations to help rebuild the areas of France and the Soviet Union destroyed during the war.

Stalin's German policy was similar to his strategic approach to Europe as a whole: in the long run, he wanted a Communist Germany —which Soviet leaders since Lenin had seen as the key to a Communist Europe—but he recognized that, in the short run, he might have to settle for a reliable regime in the territory directly under his control. Meanwhile, he wanted to exact reparations, especially from the industrial heartland now under western occupation. The preliminary solution to the German question, which was confirmed at Potsdam in the summer of 1945, appealed to Stalin because it kept both options open: Germany remained a single entity, to be governed jointly by a four-power Control Commission but administered separately in four occupation zones, each under its own military commander. Berlin, which was in the Soviet zone, replicated this situation: a collegial authority superimposed on four separate administrative sectors.

The evolution of the German problem precisely registered the progressive alienation between East and West. In 1946, the British and Americans combined their zones of occupation and refused to transfer any more reparations to the Soviets. A year later, superpower rivalries—in Greece, Turkey, and Iran, as well as in eastern Europe—paralyzed the Control Commission, leaving more and more decisions to the commanders in each occupation zone. As a result, two Germanies began to take shape, one dominated by the Soviets, the other by the three western democracies. By the end of 1947, partition had become inevitable.

In the western zones, partition was inseparable from economic recovery and political autonomy. The British and Americans, followed reluctantly by the French, recognized that European prosperity and stability would not be possible without the reconstruction of the German economy. Thus in 1948, in violation of the agreements they had made about retaining German economic unity, the western powers introduced a new currency in their zones. The Soviets responded by closing off land access to Berlin. The Allies countered with an airlift, in which American and British planes supplied the city until Stalin backed down and reopened the overland routes eleven months later. The first of a series of crises over Berlin, the blockade was a turning point in the emergence of the postwar world, not only because it solidified the division of Germany and affirmed the Allies' willingness to encourage German economic recovery, but also because it marked a dramatic shift in American opinion about its former enemy. In just three years, Americans began to see the citizens of Berlin not as willing accomplices in Hitler's crimes but as welcome allies in a new struggle against the Communist menace. In 1949, the Federal Republic of Germany was established in the West, swiftly followed by the German Democratic Republic in the East.

With the formation of the two German states, the foundations of the postwar order were in place: parliamentary democracies in the West, Communist "people's democracies" in the East, with each side linked to one of the superpowers by a network of political, economic, and security arrangements. There were, to be sure, a number of anomalies and open questions. Yugoslavia had an orthodox Communist regime but was politically independent of the Soviet Union; Finland had a democratic government but its foreign policy was dominated by Moscow; the future of Austria, still under four-power occupation, was uncertain, as was the ultimate status of Berlin, which was no longer blockaded but remained politically and militarily isolated. The superpowers hoped to expand into each other's sphere of influence, but neither was ready to use force to do so.

The keystone of the postwar order was the superpowers' sometimes perilous, occasionally precarious, and always problematic answer to the German question. That this solution would last so long was by no means clear when the two German states were founded in 1949. For more than a decade thereafter, superpower relations were strained by crises emanating in Germany, usually involving the militarily vulnerable and symbolically potent city of Berlin. As late as 1963, the Soviet foreign minister and the American secretary of state agreed that Germany represented the most serious problem in East-West relations. Even when the Cold War's epicenter shifted away from Europe, the border between the two German states was haunted by the constant danger of an armed confrontation between the United States and the Soviet Union. But the intra-German border connected as well as separated the two superpowers, joining them in a common defense of the bipolar system on which the postwar order depended.

The foundations of postwar Europe were not just military and diplomatic. From the catastrophes of the 1930s, the western powers had learned that international security and domestic stability were inseparable, the one impossible to achieve and sustain without the other. As the American secretary of state Dean Acheson told a group of newspaper editors in April 1950, "There is no longer any difference between foreign questions and domestic questions. They are all part of the same question." A successful foreign policy required what Acheson called "total diplomacy," a combination of political, economic, and military measures that could check both Communist subversion and the advance of Soviet power. From Washington's perspective, the political and economic dimensions of European security were especially important, not only because the Americans feared that Italy or France, which both had large Communist parties, might follow eastern Europe into the Soviet camp, but also because they continued to hope that an economically strong western Europe could take responsibility for its own defense.

While Washington's European allies recognized the close connection between foreign and domestic policies, they defined each according to their own national interests and experience. To the British, security and prosperity had a global as well as a European dimension; despite the dramatic decline of their power and influence, British statesmen clung to the hope that they could protect some part of their imperial heritage. But they also knew that, both economically and militarily, they were dependent on the United States. This was the painful lesson of 1940 when, after the fall of France, Britain had stood alone against Hitler, barely able to survive and totally unable to prevail without help from across the Atlantic. Memories of 1940 clearly animated a policy paper from January 1949 that urged that British support for a European recovery be shaped by a concept of "limited liability." Assistance to Europe, the paper insisted, can never leave "us too weak to be a worthwhile ally for U.S.A. if Europe collapses." The catastrophic situation of 1940, this time with the Soviet Union playing Germany's role as master of the continent, might arise again.

The memory of 1940 was also powerfully present in France, but it had a different meaning in Paris than in London. Like 1870 and 1914, 1940 confirmed the centrality of the German question, which to the French, who were unprotected by the Channel (not to mention the Atlantic), was necessarily a question of life and death. Charles de Gaulle, who had managed to elbow his way into the victors' ranks with a combination of moral authority, self-confidence, and sheer cantankerousness, had initially wanted to solve France's German problem by controlling as many of Germany's resources as his reluctant allies would allow. To do this, he was willing to remain on the sidelines as relations between the United States and the Soviet Union deteriorated: the Americans and Russians were far away, the Germans all too near. Nevertheless, de Gaulle and his advisers shared Acheson's conviction that domestic and foreign policies were now fused. French survival depended not only on containing Germany, but also on strengthening France, which would require political reform, economic development, and social modernization. The corrupt, indolent France of the Third Republic, the France of appeasement and defeat, had to be replaced by a modern, efficient nation, willing and able to lead a new Europe. Gradually, French leaders came to realize that the creation of this new France would entail a reformulation of how they viewed the German question.

Nowhere was the connection between domestic and foreign policy closer—or more painful—than in Germany. In the eastern zone, foreign and domestic affairs were dominated by the Soviets, acting through the occupation authorities and a small group of German Communists who had arrived with the Red Army. In the West, relations between Germans and the occupying powers were more complicated, but here, too, the occupiers retained the political initiative. When the West Germans were offered economic aid and greater political autonomy, they accepted even though they knew that their own recovery and independence would necessarily mean separation from their fellow Germans in the East. Konrad Adenauer, the Federal Republic's first chancellor, tried to obscure this harsh reality with talk about the "positions of strength" that would eventually lead to unification, but his domestic and foreign policies rested on a willingness to live with the fact of national partition gracefully and unambiguously.

By 1949, French policy makers realized that German recovery was unavoidable, in part because they now acknowledged the significance of a Soviet threat, in part because they saw that European recovery would be impossible without Germany's economic resources and institutions. The question then became, How can the search for French prosperity be reconciled with the search for French security?

In a radio address delivered on May 9, 1950, the French foreign minister, Robert Schuman, offered an answer to this question that would have lasting significance for the future of Europe. France, Germany, Belgium, the Netherlands, Luxembourg, and Italy, Schuman suggested, should jointly create an organization to control coal and steel production. Schuman's plan, in the words of John Gillingham, its foremost historian, "was bold, simple, imaginative and discerning—a public relations coup of heroic proportions." Despite its rather limited impact on economic development, the European Coal and Steel Community was a revolutionary event. It established the institutional template for the future European Economic Community and its successor organization, the European Union. Equally noteworthy, it expressed a radically new approach to the European order. The goal was a united Europe, to be reached through a series of small, practical steps. "Europe will not be made all at once," Schuman declared, "or according to a single plan. It will be built through concrete achievements which first create a de facto solidarity." After May 9, to quote Gillingham again, "the word 'Europe' would never be spoken in quite the same way again."

Schuman had hoped that by integrating the German economy into a European system, it might be possible to derail moves to create a German army. A month after the announcement of the Schuman Plan, however, events on the other side of the world made German rearmament unavoidable. In June 1950, North Korean troops moved south across the border established by the United States and the Soviet Union five years before, conquered much of the Korean peninsula, and threatened to destroy the outnumbered American army of occupation. Combined with the events of the preceding year, especially the victory of Communist forces in China and the successful testing of a Soviet nuclear device, the fighting in Korea signaled an ominous turn in the global conflict between the superpowers. Western leaders did not expect an immediate attack on West Germany, yet they could not ignore the similarities between the two divided and occupied regions. If Stalin was willing to encourage the use of force to control the entire Korean peninsula, might he not be tempted by the much richer prize in the center of Europe? A credible defense of Europe was essential, and, like a flourishing European economy, it was unthinkable without Germany.

In order to overcome resistance to a new German army, the United States was prepared to increase its military commitment to NATO by accepting the creation of a Supreme Allied Headquarters under an American general, who would command both American troops in Europe and NATO forces. In effect, Washington promised not only to defend Europe against a Soviet attack but also to guarantee that a rearmed Germany would not become a threat to its neighbors. In order to underscore the significance of this guarantee, Dwight Eisenhower, the most popular and prestigious soldier in the world, was made NATO's first commander. In less than a year, the United States had moved from making the sort of vague and qualified promises expressed in Article 5 of the North Atlantic Treaty to becoming structurally embedded in a European military organization.

While German rearmament was now certain, the nature of the new German army was still open. Here again the French took the lead, proposing the creation of a European Defense Community composed of divisions from France, Italy, Germany, Belgium, the Netherlands, and Luxembourg (Britain made it clear from the start that it would not participate). These national units would form a European army, which would in turn be part of NATO. The power of the European Defense Community was limited; it could neither declare war on behalf of its members nor determine strategy. Nevertheless, it represented a radical break with European political traditions: by restricting a state's right and ability to defend itself, membership in the community would undermine the foundation and instrument of sovereignty. The European Defense Community's greatest appeal was also its greatest drawback. Schuman had proclaimed proudly that "there will be German soldiers but no German army"—but the same thing, alas, would be true of France. The end of the French army was not something most politicians were willing to accept; in the summer of 1954, a clear majority of France's National Assembly effectively killed the project.

The stillbirth of the European Defense Community was one of the most contentious episodes in the crisis-filled story of European and American relations. Secretary of State John Foster Dulles, who had been one of the plan's most enthusiastic supporters, angrily warned that the French decision "obviously imposes on the United States the obligation to reappraise its foreign policies, particularly in relation to Europe." But, as it would through one crisis after another during the next four decades, the Atlantic alliance survived. Accepting a compromise proposed by the British, the Allies welcomed Germany into NATO. In return, the Germans pledged not to manufacture atomic, biological, or chemical weapons and to place their entire armed forces under NATO command. The defeat of the European Defense Community increased the importance of NATO, and therefore of the United States, as a bulwark against Communist aggression and as a solution to the problem of a rearmed Germany. The purpose of the alliance, in the British diplomat Lord Ismay's concise formulation, was to "keep the Russians out, the Americans in, and the Germans down."

As a reward for accepting an important role in the western European security community, the Federal Republic of Germany became a sovereign state. In a treaty signed on May 5, 1955, almost ten years to the day after the surrender of Hitler's Reich, the three western Allies ended their occupation and granted the Bonn regime "the full power of a sovereign state over its domestic and foreign policy." In fact, the new West Germany did not have full sovereignty. Article 24 of the treaty set limits on its domestic and foreign political autonomy by committing the Federal Republic to democratic values and integration with the West. Moreover, when they joined NATO, the West Germans had placed restrictions on themselves by promising not to manufacture certain kinds of weapons. Finally, the former occupying powers retained the right to keep troops on West German soil and in West Berlin. These limitations, which no major power would have accepted before 1914, were just one of the ways in which the theory and practice of sovereignty changed during the postwar era.

In 1955, just a month after the treaty regulating West Germany's new status was signed, representatives from the six members of the European Coal and Steel Community met in Messina, Sicily, to begin planning a new economic organization. Once again the need to control German power was the driving force. Since 1950, the West German economy had grown at a miraculous rate—industrial output, for instance, had almost doubled in five years, while foreign trade would more than triple in the course of the decade. It was clear to everyone, and especially to the French, that the institutional structure of the Coal and Steel Community was insufficient to contain this kind of economic dynamism.

On March 25, 1957, following almost two years of intense negotiations, the six states signed the Treaty of Rome, which established the European Economic Community, with its headquarters in Brussels.

From the beginning, the European Community was based on a combination of ideals and interests, including a widespread desire among Europeans to escape the poisonous rivalries of the past and careful calculations of comparative advantage by individual policy makers. When one examines the course of the negotiations that produced European institutions, it is not difficult to find the constant pull of national interest. The molders of European integration believed that it was a way of making their states stronger, better able to survive in a new and complex world. For them, Europe was, in the language of American social science, a rational choice. But it is important to recognize that hazy concepts like "interest" and "rationality" acquire concrete meaning in particular historical situations.

Only within the international order imposed by the superpowers could the intensifying economic and legal cooperation among the European states seem rational. The emergence of a new Europe, therefore, was not the cause of the long peace after 1945; peace was the new Europe's necessary precondition.

The danger of nuclear war runs through the foundations of the post-war European order like iron rods in reinforced concrete. Despite frequent crises and large-scale violence in much of the world, this danger made the United States and the Soviet Union unwilling to risk a military confrontation in Europe, the only place in the world where their armies directly confronted each other. Europe seemed to have reached the point, anticipated in such passionate detail by Ivan Bloch, at which military technology had indeed made war between great powers too dangerous and destructive to be an instrument of statecraft. The two late-born progeny of the Second World War, the atomic bomb and the ballistic missile, brought absolute war, a war without limits or restraints—something Clausewitz had considered only theoretically possible—perilously close to realization.

In the early stages of the Cold War, when the United States still enjoyed a monopoly of nuclear weapons, Europeans regarded them as a necessary response to the Soviet Union's overwhelming advantage in conventional forces. Obviously the American nuclear monopoly could not last forever. As early as April 1946, a high-level British committee pointed out Britain's vulnerability to a nuclear attack: 42 percent of Britain's population lived in cities with more than 100,000 inhabitants, all of which were within eight hundred miles of the Soviet occupation zone in Germany. Sometime between 1952 and 1956, the report concluded, the Soviets would have enough bombs to "produce collapse in this country." In fact, the Soviets' program progressed faster than most western experts expected. In 1949, they tested an atomic device; five years later, a hydrogen bomb. In 1957, when the Soviets successfully launched an earth-orbiting satellite, it was clear that they possessed intercontinental missiles capable of hitting targets in the United States.

The threat of nuclear war revealed the new European order's central tension: the continent's security depended on decisions made thousands of miles away, by American statesmen with different, and potentially divergent, values and interests. Would the United States risk a nuclear attack on its own territory in order to defend its allies? Or might the United States provoke an attack on Europe in pursuit of interests that Europeans did not share? For many Europeans, then, the threat of nuclear war was a source of both stability and anxiety, a reason to hope for peace and to fear catastrophe.

Until the late 1950s, American strategy called for an immediate nuclear response to any Soviet attack on western Europe. When Soviet nuclear capabilities increased, a number of strategists began to question the wisdom and credibility of what Secretary of State John Foster Dulles referred to as "massive retaliation." As the British general Sir John Cowley remarked in 1959, this strategy offered two bleak options: "Unless we bring the nuclear deterrent into play, we are bound to be beaten, and if we do bring it into play, we are bound to commit suicide." Didn't this paradox undermine the plausibility on which successful deterrence rested? Was it reasonable to suppose that American leaders would commit national suicide in defense of another country? But if the answer to this question was no, how else could European security be maintained?

In the early 1960s, American strategists sought to widen their range of choices by developing a policy of "flexible response," which was supposed to provide decision makers with a variety of military options — ranging from the deployment of conventional forces to the limited use of tactical nuclear weapons — that they could adopt before they were compelled to unleash strategic nuclear weapons against civilian targets. While this policy seemed to offer alternatives to what President John Kennedy once called the choice of "holocaust or humiliation," for many Europeans it raised some troubling issues.

In order to manage the escalating series of responses, a strict American monopoly over nuclear forces had to be retained, and this naturally increased Europeans' sense of dependency. Closely connected to the issue of control was the question of whether flexibility actually strengthened the West's ability to deter a Soviet attack. Since the Soviets' superiority in conventional weapons remained, might they not be tempted to seize territory and then confront the United States with the choice of accepting the situation or risking the destruction of their own society? Europeans feared that while a strategy of flexible response offered the United States the possibility of fighting a limited war with the Soviet Union, this might be done at the expense of Europe's security, perhaps its very existence. These fears were amplified by the crises that perturbed the global system during the late fifties and sixties: several great-power confrontations over Berlin, endemic instability in Asia and the Middle East, and, most serious of all, the Cuban missile crisis of October 1962, which brought the world to the brink of a nuclear exchange between the superpowers.

America's three most important allies responded to the challenge of nuclear strategy in ways that reflected their own geopolitical positions and historical experiences. The British accepted the role of junior member in a U.S.-led nuclear partnership. They had their own bombs but, after tense negotiations in Nassau, the Bahamas, at the end of 1962, adopted American-supplied Polaris submarines to deliver them. Charles de Gaulle, who had returned to power in 1958, was left out of the Nassau meetings—just as he had so often been left out of wartime discussions between Churchill and Roosevelt twenty years earlier. De Gaulle decided that France needed its own nuclear force, large enough to inflict serious damage on any potential aggressor. His hostility to the Anglo-American nuclear monopoly was one reason he took France out of NATO in 1966 and continued to seek a Europe less dependent on American political, economic, and military power.

The state with the most difficult hand to play in the western alliance's efforts to reach a consensus about nuclear strategy was, predictably enough, West Germany. Since it straddled the East-West frontier, it was obviously vulnerable to a Soviet attack. Shouldn't the West Germans, whose land was at risk and whose soldiers were so important to NATO, help make the strategic decisions on which their survival depended? Some German policy makers insisted that participation in nuclear strategy meant having nuclear weapons of their own. As Franz Josef Strauss, the defense minister and leading advocate of a West German nuclear arsenal, told an audience at Georgetown University in November 1961, these weapons are "the symbol and even the characteristic aspect of the decisive criterion of sovereignty." For several years, the United States searched unsuccessfully for some mechanism that would satisfy the West Germans' demands without allowing them to have a finger on the nuclear trigger. Against the possibility of a nuclear Germany, the Soviets exerted enormous pressure: there is strong evidence that between 1958 and 1963 Moscow provoked a number of crises over Berlin and Cuba to intimidate the western powers and thus prevent an independent West German nuclear capability. In the end, the Germans did not become a nuclear power and remained, not always comfortably, dependent on the collective power of NATO.

By the late 1960s, the security environment in Europe had achieved a kind of stability. The focus of East-West conflict had moved from Europe to the Middle East, where Israel's victory in the Six Day War of 1967 created a new set of dangers and opportunities, and to Southeast Asia, where the United States was engaged in a long and costly war. NATO survived the defection of France and successfully resisted West German efforts to acquire its own nuclear weapons. The Berlin problem, which would remain unsettled until 1989, had moved to the periphery of European affairs; the question of whether the two Germanies would ever unite had lost much of its urgency as both German states became more deeply embedded in the bipolar political landscape. The superpowers made some progress in arms control negotiations, even as they continued to expand their nuclear arsenals. There were frequent efforts at détente, interrupted by occasional crises and mutual denunciations. But each side went out of its way to avoid a military confrontation. Overall, relations between East and West had settled into that condition of hostile and competitive coexistence suggested by the very notion of a cold war. Looking back, the last two decades of the Cold War seem like a period in which a great deal happened but not much changed.

Despite the stabilization of the postwar European order, the superpowers went on deploying massive quantities of lethal hardware along the fault line between East and West. In 1986, the Communist bloc's Warsaw Pact had 94 divisions, along with 30,000 tanks and another 30,000 armored personnel carriers, stationed in East Germany; NATO had 9,000 tanks and 17,000 APCs. Both sides had thousands of tactical nuclear weapons, hundreds of thousands of tons of chemical agents, and millions of tons of conventional arms and ammunition. "Nowhere in the world," wrote the German journalist Theo Sommer, "are there so many soldiers, so much war material, and so many nuclear weapons concentrated in such a compressed area as in the two German states. If it should ever come to conflict between east and west, not much of Germany would survive." And who could suppose that if such a conflict did occur, it could be contained to Europe? Surely the chances were great that once fighting began along the German-German border, it would swiftly escalate into a nuclear exchange in which the superpowers would use the full destructive fury of their immense stockpiles of weapons.

The magnitude of potential destruction that is so coldly expressed in the data on nuclear force levels made the use of such weapons increasingly unthinkable—or, perhaps more accurately, increasingly difficult for most people to think about. The possibility of a nuclear war was like the possibility of a natural catastrophe or fatal illness. We all know that earthquakes and tornadoes, brain tumors and fatal heart attacks, can and do happen, but we usually go on living as though they won't—unless, of course, something comes along that makes us aware of the perils lurking in the future. Similarly, Europeans knew that a nuclear war might occur, either through accident or design, but most of them, most of the time, carried on with their lives as if tomorrow would be like today.

The destructive power of nuclear weapons seemed to paralyze people's political will at the same time it numbed their imagination. Considering that their national existence was at stake, nuclear issues only infrequently engaged large parts of the population. Campaigns for nuclear disarmament, while sometimes passionately intense, never attracted more than a minority of supporters. Nor were many Europeans willing to invest heavily in the conventional means of national defense that might have decreased their dependence on nuclear weapons. NATO members regularly failed to meet the military goals set by periodic efforts to reform and reinvigorate the alliance. European governments made no serious efforts at civil defense, nor were there sizable constituencies demanding that they do so. The danger of nuclear war, therefore, always shadowed but usually remained on the edge of people's consciousness, the source of a diffuse and intermittent anxiety that they had learned to accept as part of everyday life in the postwar world.

The bipolar order created by the Cold War made it possible for Europeans to live at peace with one another, but it did not bring peace to the rest of humanity. In the second half of the twentieth century, as in the period between 1871 and 1914, a relatively tranquil Europe was surrounded by a violent world. In both periods, Europeans themselves were deeply involved, but with very different results. As we saw in chapter 3, in the late nineteenth and early twentieth century, the expansion of European power had frequently provoked resistance, which European armies and their native allies then ruthlessly repressed. After 1945, the process was reversed when, in one part of the world after another, European forces were either defeated or simply gave up and withdrew. A significant element in the transformation of Europe after the Second World War, therefore, was the end of the age of European empires.

The end of empire, which has been called "the greatest transfer of power in world history," happened with remarkable speed and comprehensiveness. Between 1940 and 1980, more than eighty of the European powers' overseas possessions, inhabited by about 40 percent of the world's population, became independent. Decolonization was sometimes peaceful, sometimes bloody. In a few cases, the colonizers used massive force in an attempt to hold on to power. The Dutch sent 100,000 troops to combat nationalist rebels in Indonesia before they granted their former colony independence in 1950. The French vainly fought to reassert control over Indochina after 1945, but following their defeat by Vietnamese insurgents at Dien Bien Phu in 1954, they accepted a partition of the country, which eventually led to a second, far more destructive conflict in which the United States supported the anti-Communist South Vietnam. Often the worst violence came after independence, when competing groups struggled to fill the political vacuum left by the colonizers' departure. Such was the unhappy fate of the Indian subcontinent, where severe communal violence between Hindus and Muslims broke out in the wake of Britain's rapid withdrawal. Much the same thing occurred, with equally long-lasting consequences, after Britain's even more precipitous retreat from Palestine in 1948.

The locus classicus of empire's violent end was Algeria, where rebels waged a bitter seven-year battle against French rule. Algeria had been a French colony since 1830 and was regarded, at least in theory, as an integral part of France. It was, for example, governed by the Ministry of the Interior and was included in the territories protected by the North Atlantic alliance. Algeria had an unusually large number of European inhabitants (roughly one million, compared to eight million Arabs and Berbers), many of whom had deep roots in a country where their families had lived for generations. These settlers were firmly committed to keeping Algeria French and thereby retaining the political privileges and economic advantages they enjoyed. Equally committed to a French Algeria were members of the officer corps, who had suffered a series of humiliating defeats, most recently in Indochina, which they blamed on the republic's political, social, and moral weaknesses. Algeria, these officers believed, was the last chance to salvage their own honor and France's status as a world power.

In November 1954, six months after the fall of Dien Bien Phu, a small group of Algerian nationalists began an armed struggle for independence with a few isolated attacks on Europeans. Following a familiar spiral, terror provoked counterterror, which increased the ranks of the rebels, hardened the resistance of the settlers, and destroyed any chance of a moderate solution. By 1956, there were 400,000 French troops in Algeria, two thirds of them conscripts. At first, the overwhelming majority of the French electorate favored remaining in Algeria, but as the costs mounted, popular commitment declined. Caught between those ready to use any means to win and those in favor of a negotiated peace, the government was paralyzed, too weak to achieve victory or to accept the burden of defeat.

In May 1958, the Fourth Republic was replaced by Charles de Gaulle, who returned to power apparently determined to defeat the Algerian rebels and restore French grandeur. Three years of bitter fighting convinced him that the war could not be won. He reluctantly accepted defeat, faced down the threat of civil war in France, repressed a European counterrebellion in Algeria—and narrowly escaped several assassination attempts. After independence, most European Algerians fled, destroying as much of the country's infrastructure as they could and abandoning their numerous Muslim allies to the victors' murderous revenge.

France's defeat in Algeria was political. Militarily, the army had been able to curtail the insurgency by using torture to disrupt its urban networks and a ruthless policy of resettlement to control the countryside. The insurgents could not win militarily, but despite staggering losses, they survived; terror attacks against Europeans diminished, but did not cease. The longer the violence continued, the more obvious it became that the army and its supporters had lost the most important battle of all, the battle for public support—in the United Nations, among France's allies, and, where it mattered most, in France itself. Support for the war shrank as more and more of the French public came to regard it as morally corrupting, economically ruinous, and, in the end, unwinnable.

In retrospect, the brutal campaign to keep Algeria French seems like a struggle against the forces of history itself, an attempt to sail into what Harold Macmillan called "the winds of change" that were blowing across the non-Western world, "from the deserts of North Africa to the islands of the South Pacific." That the Algerian rebels had, from such modest beginnings, managed to defeat the formidable forces aligned against them inspired emulation throughout Africa and the Middle East. Among those who celebrated the Algerian nationalists' victory was Nelson Mandela, who saw them as models for the African National Congress. Yasser Arafat, the future leader of the Palestinian resistance to Israel, was in the cheering crowd that welcomed the rebellion's leaders when they returned to Algiers, which soon became "the Mecca of the revolutionaries."

In the course of the 1960s the winds of change carried away one European colony after another. The colonial powers were defeated in part because they had been weakened by two world wars and in part because their opponents had become better equipped, more effectively led, and more efficiently organized than ever before. But the collapse of colonialism was not simply due to military defeats like Dien Bien Phu. The Europeans lost not only the ability but also the will to retain their empires; they no longer assumed that they could and should dominate the globe. Before 1945, when Europeans suffered a military defeat, as often happened in the early stages of imperial conquest, they had kept on fighting until they eventually prevailed. After 1945, even military victories, such as the British success in Malaya in the 1950s, were followed by compromise and withdrawal. Sooner or later, with or without a fight, Europeans abandoned their empires.

Colonialism ended because of a shift in Europeans' moral calculus and, more pointedly, in their sense of what really mattered. As their subject peoples began to demand independence with ever greater insistence, the colonizers asked themselves if retaining their colonies was worth the expenditure of blood and treasure that now seemed necessary. Did it make sense to send conscripts to defend some distant territory or to spend resources abroad that could be better used at home? "Imperialism," wrote the British Labour politician John Strachey, "has ceased to bring appreciable benefits to the advanced countries (without ceasing to be ruinous for the underdeveloped)." In a few words, Strachey summarized the combination of pragmatism and guilt that undermined many Europeans' belief in their colonial mission. If colonialism did neither the mother country nor the colonies themselves any good, then there was no reason to keep it going.

The changing place of colonies in Europeans' political calculations is nicely illustrated by Charles de Gaulle. Soldier, patriot, tireless defender of French grandeur, de Gaulle was in many ways the personification of traditional statehood. Without his sure sense of France's historical identity and destiny, he would never have been able to insinuate himself into the ranks of the winners following the debacle of 1940. After the war he was convinced that the fruits of victory must include a restoration of the French Empire, the necessary foundation for France's claim to great-power status. Without its empire, he once remarked, France "might count for nothing more in the world than Greece does in Europe." Whatever de Gaulle's intentions when he returned to power in 1958, he eventually realized that a French Algeria could not be saved, at least not at a price France could afford. "Algeria costs us . . . more than she is worth," he declared in the spring of 1961. "Decolonization is our interest, and therefore our policy." De Gaulle would never have admitted that he was giving up the pursuit of national greatness. But he realized that economic prosperity, technological progress, and social modernization had superseded imperial dominion as the basis of national power in a Europe radically different from the one he had imagined in 1945.

Sometimes with reluctance, more often with relief, the European states abandoned their empires so that they could devote their energies and resources to other issues. They became, as the British historian D. K. Fieldhouse once wrote, not poorer but smaller: "They had been the centres of vast empires, now they were petty states preoccupied with parochial problems. Dominion had gone and with it the grandeur which was one of its main rewards." But by the 1960s, grandeur was no longer an important goal for European states. What mattered, as we shall see in the next chapter, was material well-being, social stability, economic growth. This is what European electorates demanded of their governments, and this is what governments struggled to provide. Colonial violence now seemed wasteful, anachronistic, and illegitimate, part of a vanished world in which the ability to wage war had been centrally important to what it meant to be a state.

8

The Rise of the Civilian State

IN THE FALL OF 1937, the American political scientist Harold Lasswell published an essay entitled "Sino-Japanese Crisis: The Garrison State Versus the Civilian State." Lasswell was responding to the outbreak of war in China that had been triggered by the clash between Japanese and Chinese forces at the Marco Polo Bridge, south of Beijing, a few months earlier. But he was also contributing to a long-standing debate about the role of war in modern society. As we saw in chapter 2, many nineteenth-century theorists had argued that military institutions would necessarily atrophy as societies modernized, commerce expanded, and power shifted from traditional warrior elites to merchants and manufacturers. Writing on the eve of the Second World War, Lasswell suggested that these theorists might well be wrong. Everywhere in the world, he argued, civilian states, run by businessmen and other specialists in bargaining and compromise, were giving way to garrison states, "in which the specialists on violence are the most powerful group in society." In these militarized states, every aspect of social life—production, administration, and culture—is directed toward war, and "all social change is translated into battle potential."

In the decade after Lasswell offered this "developmental construct" about the future of world politics, there was reason to expect that garrison states were indeed on the ascendancy. Specialists in violence, sometimes professional soldiers, sometimes at the head of militarized political movements, held power in many states. When the leaders of civilian states failed to bargain or compromise with these violent forces, they had no choice but to call upon their own violent specialists in order to survive. George Orwell's dystopia, *1984*, published in 1949, gave Lasswell's image of a state wholly organized for war its canonical literary expression. This was, many feared, what the future would look like.

As we have seen, Europe's future did not resemble either the militarized world of Lasswell's garrison state or the totalitarian agonies suffered by Orwell's unfortunate protagonist. Instead, amid the wreckage left by total war, Europeans constructed civilian states. Specialists in violence remained, but everywhere they were subordinated to experts on bargaining and compromise. These civilian states were organized for peace, not war; in them, social change was translated into economic production, not battle potential.

In the previous chapter we examined the primary reason for the rise of these civilian polities: the bipolar order imposed by the superpowers, which essentially eliminated the possibility of war among the European states. Within this order, western European states were willing and able to join supranational institutions such as the European Coal and Steel Community and the European Economic Community. Gradually, the influence of these institutions cracked the hard shell of the states' autonomy, qualifying their control over economic policy, the legal system, and many of the other aspects of public life that sovereign states had once so jealously defended from outsiders. Both institutionally and territorially, the boundaries of European states became more open and permeable than ever before.

The creation of the postwar international order was the necessary but not sufficient cause of the European states' transformation. Equally important were changes in the relationship between states and their citizens, which redefined the states' institutional structure and political purpose.

If the Cold War was the most consequential development in European international relations after 1945, the most significant domestic development was the explosive economic growth that affected every aspect of public and private life. This growth was unevenly distributed, but everywhere in western Europe it enabled states to recover rapidly from the disasters of the war. By the mid-fifties, ruined cities were rebuilt, food became plentiful, and stores were filled with a great variety of consumer products. By the end of the decade, a new kind of society had begun to emerge. To understand the full impact of this affluence, we must remember the economic dislocations of the 1930s and the devastating consequences of the Second World War. The prosperity that Europeans enjoyed in the postwar era drew part of its political and cultural power from people's memories of the sufferings they had endured during and right after the war.

By every measure, western Europe's economy expanded at an unprecedented rate in the 1950s: the manufacture of steel doubled, the production of grain, dairy goods, and meat equaled and then far surpassed prewar levels. In West Germany, real wages almost tripled between 1953 and 1973; in Britain, which was a laggard compared to its neighbors, wages nonetheless doubled over these two decades. By the mid-seventies, most households had refrigerators, washing machines, and television sets. Goods and the leisure time that had once been reserved for a small and privileged elite now spread: more people than ever before owned automobiles, traveled abroad, retired in relative comfort.

Much of this prosperity depended on commercial relations among the western European states: in 1953, a quarter of the European Community's imports came from another member, and by 1960, it was more than a third. Exports within the Community reached $100 billion by 1970, almost $300 billion by 1980, and more than $700 billion in 1990. Ludwig Erhard, the economics minister who was an architect of West Germany's economic miracle, may have been overstating the case when he said, "Foreign trade is quite simply the core and premise of our economic and social order." But there is an element of truth in Erhard's conviction that commerce was the postwar order's central impulse, both in the international institutions that tied the European Community together and in the domestic politics of the individual states.

The legitimacy of every western European government depended on its capacity to sustain growth and prosperity. Policy makers had learned a great deal from the upheavals of the interwar years as well as from their efforts to manage wartime production; they developed better instruments to plan economic expansion, control their currency, and ease the strains of uneven growth. Economic success, therefore, both reflected and affirmed the power of the state. Politicians knew that these were the issues their electorates cared about most. Postwar public opinion surveys demonstrated again and again that Europeans ranked the economy at the top of their list of political priorities. In a poll taken in 1975, for example, 85 percent of West Germans put economic issues (including energy policy) first. The American social scientist Ronald Inglehart's data measuring what European publics regarded as their state's main political goals show a similar picture: a majority of respondents in every European country mentioned stabilizing the economy, fighting inflation, and promoting economic growth as their first or second most pressing concern, while only a small number pointed to issues like expanding political participation or protecting free speech.

The public also expected their states to provide social services. In addition to education, governments now became responsible for health care, unemployment benefits, old age and disability pensions, and low-cost housing. Some of these welfare measures can be traced back to the late nineteenth century, but most of them were the product of the Second World War and its aftermath. As Lord Beveridge, the author of the report that would lay the groundwork for the British welfare state, wrote in 1943, "The general effect of the war" was to make "every able-bodied person in the community an asset." The French law on Social Security passed in October 1945 spoke of the "spirit of brotherhood and reconciliation that marks the end of the war." Increasingly, therefore, states accepted the obligation to provide what every citizen needed to be a productive, healthy, and secure member of society.

The clearest measure of these expanding obligations is the growth and distribution of public expenditures. In 1898, the British government's expenses represented about 6.5 percent of the gross domestic product; a century later, it was about 40 percent. In 1898, about 10 percent of Britain's budget was devoted to "Education, Art, and Science"; the biggest budget item, 36 percent of the total, was for defense. By 1998, the British government was spending 30 percent of its budget on Social Security, another 17 percent on health, and 12 percent on education. In Germany, which had a relatively well-developed welfare program by the end of the nineteenth century, the rise in social expenditures was not as steep as in Britain, but it was no less impressive: 40 percent in 1912, and in West Germany, 69 percent in 1950, and 73 percent in 1975. In the Netherlands, spending on social programs in 1970 ranked fourth (behind education, public works, and defense); ten years later, it was more than eight times larger and well ahead of its competitors for public funds.

These data measure not only an immense expansion in the size of the state's apparatus, but also a dramatic shift in the state's central function. At the beginning of the century, the largest expenditure of every European state was defense. As spending on social services grew, the military's share of the budget declined, sometimes precipitously. In Britain, for example, defense spending, in constant prices, remained virtually stable between 1955 and 1979 (2.58 and 2.63 million pounds), but its relative position substantially declined, from 25.1 to 10.9 percent of the government's total budget. In Holland, the decline of defense spending was comparable, from 18.3 percent of the budget in 1960 to 9.8 percent in 1980; at the same time, outlays for defense declined from 4.2 percent to 3.3 percent of the gross national product, which expanded over the same period from 41.5 to 333.9 billion guldens. In other words, while economic growth provided states with much higher revenues, none spent these additional resources on the military. As the Austrian political economist Rudolf Goldscheid once remarked, "The budget reveals the true nature of the state after its deceptive ideology has been ruthlessly stripped away." The budget of postwar European states revealed their civilian nature.

The military's declining claim on the state's resources accurately mirrored European citizens' attitudes toward security issues. For example, in 1973 Ronald Inglehart found that only a very small minority of those he surveyed put "strong defense forces" first or second among their political priorities — as few as 2 percent in Belgium and Denmark, 3 percent in France, 5 percent in Germany, and 6 percent in Britain. During the 1980s, when Margaret Thatcher's Conservative government in Britain tried to make defense a major issue, no more than 20 percent of those polled ranked it as the top political problem, and a much smaller percentage was willing to increase military spending, especially at the expense of social services. Of course these data do not show that Europeans did not care about national security. Nobody wanted another country's army violating their borders or foreign troops occupying their land. The data do suggest that because most people no longer believed that this was likely to happen, they wanted their governments to devote their attention to other, more pressing matters. Increasingly, when Europeans worried about security, they had in mind not the protection of their states from foreign enemies, but rather the protection of their own immediate interests and future well-being.

We can trace the changing character of the European state by examining the evolution of the mass conscript army after 1945. Conscription, as the British historian Victor Kiernan has written, "is always a significant index of the society where it is found; to view it solely as a method of conducting wars is to see very little of it." This was true at the beginning of the twentieth century, when Europeans regarded conscript armies as schools of nationhood, and it remained true in the century's last decades, when these armies lost their hold over people's political imagination.

Like the modern state with which it was traditionally associated, the mass conscript army survived the war. Even in Britain, where compulsory military service had been reluctantly and belatedly reintroduced in 1939, the Labour government elected in 1945 decided to retain conscription, both to meet Britain's global obligations and to persuade the United States of its political reliability and military significance. In France, the tradition of military service was strong and political opposition to a professional army deeply entrenched. According to the French national service law passed in 1950, men were required to serve eighteen months of active duty (eventually increased to thirty months) plus three years in the active reserve, sixteen in the inactive, and eight more on standby—in other words, a twenty-eight-year commitment from which there were few exemptions. Despite intense opposition from a minority and widespread apathy among the majority of the population, the West German government introduced conscription in 1956, eventually raising an army of half a million. Everywhere in Europe, therefore, states had either retained or reintroduced the draft in the immediate postwar period.

With the exception of Great Britain, where the national service army was replaced with a professional force in 1960, the other European states continued to have some form of a draft until the end of the twentieth century. But over time, the political, cultural, and military significance of these conscript armies dramatically diminished. In large measure this was because the stagnation of the military's share of government expenditures severely restricted both the size of the army and the quality of its training and equipment. The French government, for example, tried to contain costs by increasing the number of deferments (including, by 1967, about 45 percent of those eligible to serve) and by raising the physical standards for induction. In most states, the period of active duty was shortened (from an average enlistment of eighteen months to twelve) and reserve commitments were curtailed. The result was a slow contraction of the army and a sharp decline in its military effectiveness. With the end of the colonial wars and the fading possibility of a European conflict, only a few professional soldiers and strategic thinkers worried about these changes in quantity and quality. In most of Europe, the existence of conscript armies was vigorously protested by a minority, passively accepted by a majority, but enthusiastically celebrated by almost no one.

In response to public pressures, some armies relaxed discipline and lowered performance standards; a few, like the Dutch, introduced regular hours and overtime pay and allowed unionization—in short, made serving in the army a job like any other. Before 1914, conscription was supposed to infuse civil society with the virtues of discipline, patriotism, and self-sacrifice. After 1945, western European armies were increasingly influenced by the values and habits of the civilian world. The result was the opposite of militarization, a process that the military sociologist Jacques van Doorn called "civilization," by which he meant the increasing penetration of civilian habits and values into military institutions.

In order to mute public opposition to conscription, most governments made it increasingly easy to perform alternative service. After 1968, for instance, a Danish conscript could select a nonmilitary option without offering a reason. In West Germany, the number of young men performing civilian service steadily increased, from 4,000 in 1964, to 77,000 in 1988, to nearly half of those inducted by the 1990s. A majority of Germans regarded this alternative service as equally sensible and productive. Some even suggested that the experience of working in schools, hospitals, and facilities for the elderly or disabled was the real school for the nation, where young men learned the caring roles often classified as feminine. These were the skills a civilian society required.

Throughout the nineteenth and first half of the twentieth centuries, the duty to fight and perhaps to die for the state was the (male) citizen's sacred obligation; the right to impose this obligation was the main source and expression of the state's legitimacy. This obligation was at the core of what the French historian Ernest Renan had in mind when he defined the nation in 1882 as "a great community constituted by past and future sacrifices." After 1945, compulsory military service remained, but it lost its association with killing and dying and thus much of its practical and symbolic power to represent and create the national community. By the end of the twentieth century, Michael Howard noted, "death was no longer seen as being part of the social contract." A few months or a year in the army was an interruption in a young man's ordinary life, welcome for some, annoying for others; it was no longer an experience infused with the ideals of heroism and emotionally charged by the possibility of sacrifice.

It is easy to overlook the demilitarization of European society because it was the result of an almost invisible revolution. The ideological confrontation of pacifism and militarism, which had intensified at the end of the nineteenth century and then dominated political discourse between the wars, had largely disappeared. Indeed, it is remarkable how little space these issues were given in the scholarly works that shaped contemporary views of European society and politics in the 1960s and 1970s. Military values and institutions, like the conscript army itself, faded away so gradually that few people noticed that they were no longer there.

To make this invisible revolution visible, we need only compare the role of the military in European society at the beginning and at the end of the twentieth century. In 1900, every European city was full of men in uniform, its streets the scene of martial rituals and the monumental reminders of past victories. By 2000, the military's role in the symbols and ceremonies of statehood had everywhere diminished. Men in uniform were rarely seen in public, and the uni-

forms themselves had become self-consciously colorless and nonde-
script, more like those of mail carriers and bus drivers than like the
splendid garb of the prewar era. At the beginning of the century, the
officer corps was associated with the aristocracy, its members prominent at court and much sought after as dinner guests and prospec-
tive bridegrooms. After 1945, the social status of professional soldiers
declined steadily. In France, for example, surgeons were put at the
top of the list by 58 percent of those polled in the mid-seventies,
high-ranking officers by only 6 percent.

In some states, to be sure, residues of the past remain. In France
and Britain, military parades are still part of the national celebrations
of Bastille Day or the queen's birthday; colorfully dressed troops still
welcome official visitors to Paris and stand guard in front of Buck-
ingham Palace. But even in these states, the cultural power and political salience of military symbols have clearly ebbed; they have become tourist attractions rather than ways to stir patriotic emotions.
In Berlin or Rome or Vienna, the public display of military power
is rare, to say the least. Most public buildings are guarded by police
officers, not sentries in dress uniforms.

Not long ago, London's lord mayor suggested that the military
monuments should be removed from Trafalgar Square. This did not
happen, but it is highly unlikely that new ones will be added. When
was the last time a European city built a monument to a military
hero or named a street after a battle? In Berlin, the Defense Ministry faces a street named for a leader of the naval mutiny in 1918 and
is flanked by one named for Claus von Stauffenberg, a martyr of
the German resistance to Hitler, and by one named for Hiroshima.
Christian Daniel Rauch's statue of Frederick the Great still stands
on Unter den Linden, but the newest monuments in the German
capital are dedicated to war's victims, not its heroes.

Violence diminished but did not disappear from European public
life in the postwar era. There was some bloodletting in the immediate aftermath of the Second World War: acts of vengeance against
collaborators, skirmishes between Communist and non-Communist
forces in Poland, anti-Soviet resistance in Ukraine, fighting between
Albanians and Serbs in Kosovo, and—most tragically of all—po-
groms against Polish Jews trying to return home from the death

camps. After 1945, however, there was no massive wave of social and political violence comparable to the revolutionary upheavals and civil conflicts that had raged after 1918. Only in Greece was there a full-scale internal war to determine the nation's future. Elsewhere, the forces of order, backed by the formidable armed might of the super-powers, quickly imposed their will on an exhausted and demoralized population. Within a short period of time, violence was restricted to the periphery, places like Ireland, Spain, and Cyprus, where ethnic and religious rivalries could still turn murderous.

In most of Europe, the overwhelming majority of people came to view violence, both domestic and international, as something to be feared and avoided, not applauded or excused. War and revolution no longer seemed, as they had to many in the nineteenth century, like the "engines of history," the essential sources of progress or the inevitable expressions of humanity's struggle for existence. Nor was violence celebrated as it had been in the years between the wars. Radical right-wing movements, the chief institutional and ideological support for political violence in general and military values in particular, seemed to have been terminally discredited by their association with the horrors of Nazism. Everywhere these movements were banned or simply disappeared from the political scene. It seemed as if the experiences of the twentieth century had finally taught Europeans that such turmoil was an aberration, a pathological assault on normal society, something to be combated and over-come, like crime. Those who wanted to shed blood, for whatever reason, were now regarded as criminals, fanatics, or maniacs, not idealists, heroes, or saviors.

In the late 1960s, some people feared that political violence might be coming back. Reporting what should have been apparent to any-one who had not been traveling in outer space, the Central Intelligence Agency informed President Lyndon Johnson in September 1968, "Dissidence, involving students and non-students alike, is a world-wide phenomenon." As in 1848, when revolutionary fevers seemed to infect all of Europe, in 1968 "dissidence" was an international movement. Although their methods and goals varied from place to place, protesters everywhere drew on three sources of discontent. The first was the war in Vietnam. A pressing political issue (with serious practical implications for young men of military age) in the United States, Vietnam had great symbolic power elsewhere, since it brought into focus anti-Americanism, antimilitarism, and sympathy for Third World peoples. The second source of dissent was a diverse and somewhat contradictory set of discontents that joined those who rejected the materialism of the postwar economic boom with those who believed they did not get their fair share of its benefits. Finally, and perhaps most important, was a cluster of generational experiences that gave young people from Berkeley to Berlin a sense of solidarity, expressed in the songs they sang, the clothes they wore, and the revolutionary heroes they revered. The generation of 1968 was the first postwar generation, the first cohort of Europeans who, because they did not have direct experience of depression and war, could focus their attention on society's present failings rather than on its impressive progress over the past.

The would-be revolutionaries who took to the streets in 1968 believed they were participating in a new production of the great drama that had had its first performance in 1789. This identification with a revolutionary lineage was, of course, especially true in France, where the events of May 1968 seemed to make Paris once again the center of an international uprising against tyranny and injustice. Throughout the Western world, young people built barricades, went without sleep to hear interminable speeches, and clashed with the forces of order. Everywhere charismatic leaders emerged to articulate political demands and inspire enthusiastic crowds. Everywhere the tiresome habits of everyday life were broken by the thrilling promise of revolutionary action.

However convincing an imitation of revolution these protests may have been—for a brief time even Charles de Gaulle believed that his regime was in danger—1968 was not 1789 or, for that matter, 1830, 1848, or 1917. No one was forced from office by these protests, society was not transformed, political institutions were shaken but not overturned. Moreover, in comparison with revolutionary moments of the past, the protests of the late sixties were remarkably peaceful. Of the 44,681 violent deaths in France during 1968, only 6 were politically connected; almost all the rest were the result of road accidents, violent crimes, or suicide. A careful study of some 1,767 "protest events" in West Germany between 1965 and 1989 found that 80 percent involved no violence at all and fewer than 2 percent produced serious injuries. On occasion, security forces could be brutal, even lethally so, but in general the civilian states of western Europe defended themselves without the use of deadly force.

By 1969, popular support for the protest movements had begun to ebb. Most of the protesters went back to their ordinary lives, while some of their leaders began a long process of reconciliation with the established order, which in a few cases would eventually lead to a government post or even a ministerial portfolio. A small minority of dissenters, however, shifted from popular agitation to conspiracy, from rallies and demonstrations to kidnapping and murder.

Political terrorism was especially murderous in Italy, in part because of the government's corrupt incompetence, in part because terrorist groups emerged on both the extreme right and left, combining to produce a sense of systemic political crisis. From 1969 to 1980, political violence in Italy killed 415 people and wounded more than 1,000 others. In 1978, the kidnapping and murder of a former prime minister, Aldo Moro, by the Red Brigade finally compelled the government to take decisive action. Over the next few years, the leaders of the radical leftist group were apprehended and its effectiveness declined.

In the Federal Republic of Germany, the terrorists were fewer in number but equally ruthless and indiscriminate in their use of violence. Materially supported by the German Democratic Republic in the East, the Red Army Faction was morally sustained by the conviction that the Federal Republic was a fascist regime against which any means of opposition were justified. That it was, in fact, a constitutional welfare state, led by Willy Brandt, a Social Democratic veteran of the anti-Nazi resistance, seemed to make little difference. German terrorists carried out more than 200 bombings, 69 bank robberies, and many kidnappings; they murdered 28 people. By the late seventies, however, their leaders were either dead or in prison. Although remnants of the organization survived underground into the 1990s, revolutionary radicalism had clearly lost its energy and appeal.

Terrorists in West Germany and Italy remained on the margins; they were a handful of fanatics who had few active supporters and were admired (usually from a safe distance) by a small minority of the population. In Northern Ireland, by contrast, the terrorists were able to acquire a much broader social base by mobilizing deeply rooted communal antagonisms. Beginning in 1967, the Catholic minority, inspired by protest movements in the United States and Europe, began to agitate for equal rights and opportunities. Their efforts, which were initially nonviolent, provoked harsh countermeasures from the Protestant-dominated security forces as well as from militant Protestant organizations.

As often happens, violence became reciprocally reinforcing when each side responded to the other's outrages. The Irish Republican Army, which had been dormant since the end of the Irish Civil War in the 1920s, became active again; it was determined to drive out the British once and for all. In 1969, the British troops who had been sent to restore order were swiftly drawn into the conflict. At the high point of hostilities in 1972, there were 467 victims of the shootings and bombings carried out by both sides. For the next several years, fatalities ranged between 200 and 300, before declining to fewer than 100 annually.

In prosperous, well-organized societies terrorism plays a paradoxical role. On the one hand, they are highly vulnerable to a minority willing to use violence. If the terrorists can count on a few sympathizers, they are able to use the mobility and anonymity inherent in these societies to elude capture. Moreover, modern societies are full of potential targets—prominent people who can be kidnapped, monuments that can be bombed, members of the security forces who can be assassinated. On the other hand, these societies are remarkably resilient. Their social and political fabric may be strained, but it will not be shredded by terrorist attacks. Eventually, Europeans learned that kidnappings, bombings, and assassinations may produce outrage and encourage a sense of crisis, but they will not bring down the social and political order. The terrorism that emerged in western Europe during the seventies either gradually dissipated, as it did in West Germany and Italy, or, as in Northern Ireland, was reduced to what the British home secretary called "an acceptable level of violence." In practical terms, violence remains at an acceptable level when it does not interfere with most people's daily lives. The unrest of the late sixties and seventies, therefore, tested but did not destroy the postwar civilian state.

In the 1970s, the most significant political changes in Europe were not produced by pseudo-revolutionary violence but by a remarkably peaceful process that transformed authoritarian regimes in Greece, Portugal, and Spain into stable liberal democracies.

In 1949, Greece emerged from its bitter civil war economically weak, politically divided, and spiritually exhausted. Persistently assaulted from left and right, democratic institutions were further undermined by a series of scandals involving leading politicians. In April 1967, a military coup established a squalid, unpopular regime that imperfectly concealed its reliance on terror and intimidation behind the threadbare veil of anticommunism. In 1973, the chief of the military police, General Demetrios Ioannidis, formed a new, equally brutal, but singularly incompetent government. In July 1974, Ioannidis orchestrated a coup on Cyprus, which provoked a Turkish invasion and occupation of the northern part of the island to protect the interests of the Turkish minority. It seemed for a time that Greece and Turkey, both members of NATO, might go to war, but the crisis ended when civilian control was restored under the veteran parliamentarian Konstantinos Karamanlis, whose policies were ratified by a great election victory in November. In 1975, a coup attempt by dissident army officers failed miserably; it was clear that the era of oppressive military rule had come to an end.

In April 1974, three months before the Greek junta was sent packing, a group of Portuguese army officers, disenchanted with the apparently endless colonial wars in Africa, installed a new government under General Antonio Spinola. After being compelled to preside over radical reforms of which he did not fully approve, Spinola attempted a countercoup. The situation remained unstable, with threats from both right- and left-wing forces, until a moderate group of officers led by General Antonio Eanes seized power and held new elections. Eanes was elected president, and he appointed Mario Soares, a moderate socialist, as prime minister. Bumpy as Portugal's journey to democracy sometimes was, the new regime's accomplishments were impressive. In two short years, the reformers managed to end an authoritarian regime that had been in power for more than fifty years, liquidate a series of bloody colonial wars, and survive a coup by right-wing defenders of the old order—without falling into the hands of the antidemocratic left. At the same time, they installed a popularly elected government, which retained its ties to NATO and developed a strong and increasingly profitable relationship with the European Community.

Certainly the most impressive example of a peaceful transition to democracy was Spain. In the late 1960s, as Francisco Franco's dictatorship entered its fourth decade, the struggle over the nation's future intensified. When Franco began to weaken physically and mentally, many feared that a new round of civil strife was about to begin. Moderate supporters of the regime recognized that since Francoism without Franco was neither possible nor desirable, reform was unavoidable. But the reformers faced a number of powerful opponents, among whom the most formidable were Basque separatists, who responded to Franco's decline by increasing their violent attacks on the state, and the diehard defenders of the status quo, who used the resurgence of Basque terrorism as an argument against any kind of change.

The reformers had the great good fortune of having the support of King Juan Carlos, Franco's chosen successor, who realized that a peaceful and prosperous Spain had to be based on stable democratic institutions. The king backed a set of constitutional reforms that went too far for some conservatives, not far enough for some liberals, but which a majority accepted in order to avoid violence. Franco died in November 1975; two years later, national elections ratified the transformation of his dictatorship into a parliamentary democracy. The ignominious collapse of an attempted military coup in 1981 showed that a new chapter in Spanish history had truly begun.

Against the backdrop of their turbulent and often violent histories, the peaceful transition to democracy on Europe's periphery— in Greece, Portugal, and Spain—took many observers by surprise. ("When the revolution occurred in Portugal," remarked the CIA's station chief in London, "the United States was out to lunch.") In none of the three cases was the process inevitable: the role of a few personalities—Karamanlis in Greece, Eanes and Soares in Portugal, King Juan Carlos in Spain—was critical; had they failed to intervene in defense of democratic institutions, things might have turned out quite differently. But democratization was also the product of deep social, economic, and political changes. All three countries had experienced rapid economic development during the preceding decades; in the 1960s, only Japan enjoyed a higher growth rate

than Spain. Economic expansion was connected to a massive movement of Greek, Portuguese, and Spanish workers abroad, mostly to
western Europe, and an equally massive influx of foreign tourists;
both movements created social, economic, and cultural ties to the
burgeoning consumer societies of the West. Finally, the European
Community provided essential support for reform by condemning
repression, encouraging democracy—and holding out the possibility of membership. The long and often arduous process of joining
the Community, initiated by the reformers in all three countries, became an essential element in the stabilization of democracy. Equally
important, the image of a stable, prosperous, and democratic Europe, the Europe of the common market and the civilian state, provided a model for those in search of a way to transform outmoded
and ineffective authoritarian regimes.

In Greece, Portugal, and Spain, democracy triumphed peacefully
because the overwhelming majority of their citizens, even among
those who were skeptical about reform, no longer believed that violence was a legitimate political weapon. Even when they had the
means to resist change, the opponents of democracy usually lacked
the conviction necessary to die or kill. "Nations," said General Spinola in the midst of the Portuguese revolution, "prefer to live prosaically rather than to disappear in glory." The political culture of
postwar Europe was indeed prosaic—a culture of peace rather than
war, life rather death, everyday survival rather than glorious sacrifice. The same historical developments that promoted the growth of
civilian states in Europe's prosperous core encouraged the peaceful
emergence of democratic regimes along the periphery, first in Portugal, Spain, and Greece, and then, even more remarkably, in the Soviet Empire.

Thus far, our account of the postwar world has emphasized western Europe, where the eclipse of violence radically changed the relations among and within states. In many ways, eastern Europe does
not seem to fit into this story. Armed conflict was always close to
the surface of public life in Communist regimes. On three occasions—in East Germany in 1953, Hungary in 1956, and Czechoslovakia in 1968—Soviet troops were required to put down rebel-
lions. Moreover, eastern European regimes were not demilitarized

like their counterparts in the West; they continued to maintain large armies, which remained "schools of the nation," where young men were supposed to learn the doctrines and disciplines required to defend themselves against the enemies of the working class. Nevertheless, war disappeared from eastern as well as western Europe. Here, too, the bipolar Cold War order created a security system in which the possibilities of intrastate violence became increasingly remote. As a result, civilian values and attitudes slowly and unevenly developed in the East, where they established the foundation for the great peaceful revolutions of 1989.

Communist rule in Europe ended peacefully, but it began in a wave of political terror, mass deportations, show trials of alleged enemies of the regime, confiscation of farms, and forced industrialization. In the years immediately after 1945, 200,000 were arrested in Hungary, 136,000 in Czechoslovakia, 180,000 in Romania, and in Albania, with a population of just over a million, a remarkable 80,000. Compared to the high point of Stalinist terror in the 1930s, a relatively small number of these people were killed; most were imprisoned or swallowed up by the vast complex of forced-labor camps that were spread across eastern Russia. In late 1952, reports of a plot to kill Stalin, involving the Kremlin's doctors, most of them Jews, raised fears that the Soviet regime would, as it had at the peak of Stalinist terror, once again begin to devour its own progeny.

Stalin's sudden death from a stroke in March 1953 was followed by a marked shift in the Soviet system. Terror became less prevalent and less lethal as the regime adopted a policy of what the philosopher Tzvetan Todorov called "repression without annihilation." Tens of thousands of political prisoners were released (including the wife of Foreign Minister Molotov, who had been imprisoned in 1949); talk of the doctors' plot ceased; Lavrenti Beria, who had run Stalin's terror apparatus since 1938, was arrested, found guilty by a special tribunal, and immediately executed. Beria was the last front-rank Soviet leader to die as the result of an internal power struggle. In the post-Stalin era repressive measures became more predictable, less likely to strike at those close to the core of the regime, and less indiscriminately murderous.

Yet even with changes of such magnitude, life in what the Polish dissident Adam Michnik once called "post-totalitarian" society remained severely restricted. Pressures for political conformity were still in place, the omnipresent security apparatus was intact, the forced-labor camps stayed open. Dissent had a high price; success demanded disciplined obedience. In 1956, when Communist leaders in Poland and Hungary overreacted to the apparent easing of Stalinist terror, the Soviets responded vigorously. In Hungary there was a full-scale civil war, in which thousands died and tens of thousands fled the country. That year was the first but not the last time that the Soviet Empire provided evidence for Tocqueville's famous remark that attempts at reform were especially dangerous for tyrannical regimes.

Like its domestic policy, Soviet foreign policy changed after 1953, but also within well-defined limits. Nikita Khrushchev, who eventually won the struggle for power after Stalin's death, realized that the destructiveness of nuclear weapons had transformed the nature of war. An armed clash with the West, he now maintained, was no longer inevitable; communism and capitalism could, at least for a time, coexist. "Peaceful coexistence" was officially defined in 1961 as "a specific form of the international class struggle," which would continue until communism eventually triumphed. In the meantime, neither side could afford to fight a major war in which there would be no winners. Coexistence did not, of course, prevent Khrushchev from testing the West—for example, by threatening its position in Berlin or by stationing missiles in Cuba. Nor did it preclude the maintenance of a massive military establishment, sustained by what one expert called "a permanent wartime economy" that absorbed immense amounts of the Soviet Union's scarce resources.

Stalin's heirs did not abandon the essential elements of the regime that Lenin established in 1917. The party still monopolized political power. The international class struggle remained the driving force behind Soviet foreign policy. Nevertheless, the attenuation of terror at home and the possibility of peaceful coexistence with the capitalist West suggested that even in the Soviet Empire the legitimacy of political violence was waning. Beneath the hard crust of Communist rule, we can find some of the same trends at work that were transforming western European politics and society in the twentieth century's second half.

In the East as in the West, the decline of violence was matched by a corresponding increase in the importance of the economy. Khrushchev and his colleagues believed that they could govern without terror at home and triumph without war internationally because their economy could outperform its rivals. In the late 1950s and 1960s, this faith in the economic advantages of communism did not seem as far-fetched as it does in retrospect. Under Khrushchev, the Soviet economy grew at a robust rate, agricultural production improved, consumer goods became more available. Soviet citizens began to acquire television sets and refrigerators, and with them the hope for a higher standard of living, free from the hardships they had borne for so long. The Russians' claims to superiority were greatly enhanced when, in October 1957, they launched a spacecraft that orbited the earth. Could it be that, as Khrushchev flamboyantly proclaimed, the Soviet Union would "bury" its capitalist adversaries, not with a military victory but by winning the economic contest in which the West had once seemed invincible? Or, as some social scientists began to predict, would Communist and capitalist societies steadily grow more alike—the former increasingly subject to market forces, the latter to government planning—until they converged in a happy middle state that blended the best of both systems?

Despite Khrushchev's extravagant claims, the Soviet system did not surpass the West. Instead, economic growth slowed, the production of consumer goods rarely met demand, and the gap between living standards on either side of the East-West divide widened. With their promises of greater prosperity largely unfulfilled, Communist regimes continued to rely on ideological appeals and enforced conformity.

In 1968, at the same time popular protests spread across western Europe and the United States, pressure for change erupted in the East. The most vigorous reform movement was in Czechoslovakia, where a new regime under Alexander Dubček tried to revive flagging support for communism with a series of reforms. This frightened his country's more orthodox neighbors and finally provoked an invasion by Warsaw Pact troops. Dubček's attempt to create "socialism with a human face" failed; he himself was exiled to an obscure post far from the center of power. To some, the Soviets' ruthless suppression of Czech reform suggested their readiness to resist any attempt to reform communism. Those closer to events recognized that the use of force was an indication of weakness rather than strength. Yuri Andropov, who was then the head of the Soviet secret police and would eventually become premier, later conceded that the repression of Czechoslovakia was the first sign of a structural crisis in the system.

The invasion of Czechoslovakia did not derail efforts by western European governments to find new avenues of cooperation with the Soviet Union. Michel Debré, the French foreign minister, referred to the invasion as a "minor mishap" (*"incident de parcours"*) on the road to détente—which indeed it may have been for the French, if not the Czechs. Unlike the United States, whose adherence to a policy of détente with the Soviets was always qualified, Europeans saw détente as the natural expression and reinforcement of their own, increasingly demilitarized foreign policy. This foreign policy rested on commerce and economic assistance—that is, on the application to international relations of the priorities that shaped western Europe's own common life. The same faith in the transformative power of economic growth—and the same set of economic interests—that had animated the formation of the western European Community encouraged closer relations with the East. In the short run, this might strengthen repressive regimes, but in the long run, everyone in both East and West would benefit. And while they waited for this to happen, Europeans had a better chance of living in peace.

The Soviet leaders' interest in détente was also primarily economic, since they needed western loans and subsidies to keep their regimes afloat. Charting a self-defeating course much like the one Karl Marx had predicted capitalist economies would follow in times of crisis, the Soviets and their eastern European allies tried to survive by adopting policies that exacerbated their structural weaknesses. The more western loans they accepted and the more dependent they became on foreign currency, the more they whetted their citizens' appetite for the goods their own economies were unable to provide. As a result, both the elite and the ordinary people began to lose faith in their system's values and performance. Even in the 1980s, while the promise of détente faded and superpower relations deteriorated, the Communist governments' need for foreign loans continued unchecked. Between 1975 and 1985, East Germany's convertible currency indebtedness grew from $3.5 to $6.7 billion, the Soviet Union's from $7.4 to $12.1 billion, and Poland's from $7.7 to $27.7 billion. Like a tiring boxer in the closing rounds of a brutal match,

Communist regimes tried desperately to stay on their feet by cling-
ing to their capitalist opponents.

The growing crisis within communism was first apparent on the
western periphery of the Soviet Empire, where opposition movements developed in the late seventies. Just like the leaders of the
peaceful transitions to democracy in Greece, Portugal, and Spain,
reformers in Hungary, Poland, and Czechoslovakia were drawn to
the promise of western European peace and prosperity. Here, too,
dissidents hoped to effect change without resorting to violence. "To
believe in overthrowing the dictatorship of the party by revolution,"
Adam Michnik wrote in 1976, "is both unrealistic and dangerous."
Courageous men and women like Michnik challenged their governments not with bombs and barricades, but with declarations, pamphlets, and peaceful demonstrations, as well as with whimsical, car-
nival-like happenings that sought to reveal the regime's hypocrisies
and absurdities. Above all, the opposition tried, in Václav Havel's
elegant phrase, "to live within the truth"—that is, to refuse to affirm
their governments' systematic deception and moral emptiness. To
confront the state violently would have revealed its power; living
within the truth revealed its weakness.

While dissidents sometimes paid dearly for their nonviolent protests, the opposition's self-imposed restraint both presupposed and
encouraged a degree of restraint on the part of the regime. At the
height of Stalinism, living within the truth would have been useless, since no witness to the truth would have been left alive. By the
1970s, the regime barred dissenters from the professions, interfered
with their movements, and sometimes imprisoned them; it did not
murder them. This opened a narrow space in which nonconformity
could survive. Even in Poland, where dissent was most vigorous, the
authorities defended themselves with as little lethal force as possible. Unlike 1956 or 1968, the Soviet Union decided not to intervene
when the opposition, which had coalesced around the trade union
Solidarity, seemed to get out of hand. Instead, in 1981 a Polish general imposed martial law to repress what everyone knew was an au-
thentic working-class movement. For the self-proclaimed representatives of the proletariat, there could be no clearer sign of ideological
bankruptcy.

In February 1985, the *Economist* published an article marking the
fortieth anniversary of the Yalta Conference, which has often been

seen as a key date in the emerging division of postwar Europe. The past forty years, the magazine declared, were a "historically abnormal period of paralysis." Of the next forty, it asked, "What are Europe's chances of being brought to life again?" The *Economist* suggested four possible answers, of which "the first, the dullest, but also alas the likeliest, is that nothing much will have changed by the year 2025."

It is not hard to understand the editors' skepticism about the possibilities of change. At the beginning of 1985, most of the leaders of the Soviet bloc were over seventy, and many of them had been around for decades; seventy-three-year-old Todor Zhivkov, for instance, had ruled Bulgaria since 1954. Konstantin Chernenko, who had succeeded Yuri Andropov as the Soviet leader in 1984, was known to be seriously ill and had the habit of dropping out of sight for prolonged periods. Stubborn, unimaginative apparatchiks like Zhivkov and Chernenko, János Kádár of Hungary, Gustáv Husák of Czechoslovakia, and Erich Honecker of East Germany personified the ingrained resistance to change in the Soviet Empire. No wonder that many experts believed it might well last another forty years.

Although few recognized it at the time, six weeks after the *Economist* published its bleak forecast the transformation began. Chernenko died on March 10, 1985; the next day, Mikhail Gorbachev was named general secretary of the Central Committee of the Communist Party. Born in 1931, Gorbachev was the first leader of the Soviet Union who had not been actively involved in the Second World War. His rise in the party hierarchy was remarkably rapid. By 1980 he was a member of the Politburo and closely allied with Andropov, who, during his brief tenure as general secretary, had made some hesitant efforts at reform. Gorbachev realized that the time for hesitation was past: faced with a bewildering array of problems—restless populations in eastern Europe, a stagnant economy incapable of producing the goods that people demanded or supporting a military establishment that could meet the challenge of the West, and a failed war in Afghanistan—he knew fundamental reform was unavoidable. Like his counterparts in the West, he believed that such reform meant the creation of an economy that could provide prosperity, welfare, and security. With such an economy, Communist rule could survive.

Gorbachev had no intention of dismantling the Soviet system; he believed in the party's monopoly of power and in communism's historical destiny. To justify his policies, Gorbachev, like all of his predecessors, invoked the authority of Lenin, who, against intense opposition from all sides, had insisted on signing the Treaty of Brest-Litovsk, making peace with Germany. "Only later," Gorbachev wrote, "was it easy to say confidently and unambiguously that Lenin was right. And right he was, because he was looking far ahead; he did not put what was transitory above what was essential. The Revolution was saved." Like Lenin's treaty with the Germans, glasnost and perestroika, Gorbachev's new policies of openness and restructuring, were a temporary retreat necessary to save the revolution. This time, though, what began as a tactical fallback ended in unconditional surrender.

While searching for ways to overcome the fierce resistance to his domestic policies, Gorbachev embarked on a series of bold initiatives in foreign affairs. He offered radically new disarmament proposals, began to liquidate the war that Soviet forces were fighting to support the Communist regime in Afghanistan, and, most dramatic of all, suggested that the Soviet Union would no longer impose conformity on its eastern European allies. In an address to the United Nations on December 7, 1988, Gorbachev made two extraordinary concessions. First, he declared that "force and the threat of force ... should not be instruments of foreign policy." This meant people were free to choose their own form of government, a universal principle "to which there should be no exceptions." Second, he proclaimed his willingness to live with, and learn from, the rich variety of political systems that existed throughout the world. International politics, he went on, was entering a new stage that demanded "the de-ideologization of interstate relations." The ideologically driven fusion of foreign and domestic policy that had characterized Communist states' relationships with friend and foe since 1917 had come to an end.

Gorbachev's words found their loudest resonance in the Communist states of eastern Europe. In 1988, Estonia dared to proclaim its sovereign autonomy. Elections in Poland in June 1989 produced a stunning defeat for Communist candidates. Demonstrations in Czechoslovakia and Hungary forced concessions, then reform, and finally a turnover of regimes. Unlike 1956 and 1968, when, after encouraging some reforms, the Soviet Union had intervened to stop things from going too far, this time dissidents were allowed to articulate and then escalate their demands. In East Germany, 400,000 Soviet troops stayed in their barracks and watched as Moscow's most resolute and strategically important ally disintegrated, caught between popular protests and a massive wave of migrants moving west. By the turn of the year 1989–90, the Cold War solution to the German question, which had been the keystone of the European order, was disintegrating.

The end of the Soviet Empire reminds us of the process that ended Europeans' overseas empires forty years earlier. Like the colonial powers, the Soviets abandoned their dominions because they had other priorities; empire no longer seemed worth the cost; other goals were more important. But however similar their calculations, the end of empire had very different results in East and West. The western European powers swiftly recovered from the loss of their overseas territories; in fact, without their empires they were able to build more prosperous and successful societies. Even in France and Portugal, where decolonization was attended by political crisis, the final result was the creation of more stable, effective governments. In the Soviet Union, however, the dissolution of empire was part of a terminal crisis that ended in total collapse. The forces of democracy and self-determination that Gorbachev released in eastern Europe quickly spilled back across the Soviets' own borders. Between March 1990 and December 1991, fourteen Soviet republics declared their independence and established themselves as separate, sovereign states. Against this disintegration Gorbachev struggled in vain. In August 1991, a halfhearted, wholly incompetent coup — much like the attempted coups that failed to stop reform in Greece, Portugal, and Spain — uncovered both Gorbachev's own weakness and the absence of any viable alternative to reform. On December 25, the Soviet Union, for which so much blood had been shed over the years, peacefully dissolved.

In a breathtakingly brief period of time, the leaders of the Soviet Union and its closest partners, still in control of massive military forces and an enormous repressive apparatus, simply gave up. That a political transformation of this magnitude could happen so quickly

*A revolution without violence. Unarmed soldiers and
demonstrators at the Berlin Wall, November 1989.*

is rare; that such a transformation could occur with so little bloodshed has no historical precedent.

The story of the Soviet experiment from 1917 to 1991 underscores
the importance of individuals in history. Without Lenin's ruthless
will to power, his small band of radical revolutionaries would never
have been able to seize control of the Russian state; without Gorbachev's stubborn conviction that reform was essential, it seems unlikely that the end of communism would have come as swiftly or
as peacefully as it did. But while these two men made history, neither made the sort of history he expected to make. At both ends of
the Soviet story stands a monumental miscalculation, soon followed
by a most unpleasant surprise. Lenin seized power because he believed that a European revolution was imminent. It was not. Gorbachev introduced reforms because he believed that they could save
the Communist regime. They could not.

And here we come to the essential difference between the founder
of the Soviet regime and its liquidator. When Lenin's illusion dissolved, he did everything possible to stay in power. When Gorbachev saw that his policies had hastened the demise of the system

he wanted to save, he let events take their course. This difference is rooted in the character and experience of the two men and, even more deeply, in the historical situation each confronted. After taking power violently, Lenin could not let it go. Surrounded by enemies at home and abroad, he knew that he would not survive defeat. Gorbachev could afford to back down because he was confident that neither his foreign enemies nor his domestic opponents would kill him. The birth of the Bolshevik regime would have been impossible without the intense violence that had infested Europe in 1917. Its remarkably peaceful collapse presupposed the decline of violence that, by the 1980s, had transformed international and domestic politics throughout Europe.

9

★ ★ ★

Why Europe Will Not Become
a Superpower

O NE SATURDAY AFTERNOON early in 1990, a small but noisy
group of demonstrators marched down the Königsallee, a
usually tranquil street in the Berlin district of Grunewald. When an
American visitor asked what was going on, his German colleague
replied with a shrug, "Something Yugoslavian." They both went
back to work. Whatever these people were demonstrating about did
not seem important, especially when compared to the momentous
events that had filled the weeks since the fall of the Berlin Wall in
November 1989. German elections were soon to be held, difficult
negotiations among the Allies had to be conducted, the future of the
armies that had occupied Germany since 1945 was still to be decided.
Yugoslavia was far away, a minor player in the great peaceful revolution that was transforming Europe. Within a few months, however,
things Yugoslavian would become harder and harder to ignore.

We now know that by the beginning of 1990, the Socialist Federal
Republic of Yugoslavia was coming unglued. In the states of Serbia
and Croatia, recent elections had confirmed the power of Slobodan
Milosević and Franjo Tudjman, who were committed to destroying
the federal structure that Tito had created after the Second World
War. As this structure imploded, reports of communal violence became more vivid and intrusive. In the images of mass murder, ethnic cleansing, and systematic rape that flashed across their television
screens, Europeans saw victims who looked like themselves, people

dressed in jeans and running shoes, their few possessions stuffed into shopping bags carrying familiar labels. The Dalmatian coast, scene of some of the worst atrocities, had been a popular vacation spot for Europeans, who recalled with pleasure its spectacular scenery, sunny beaches, and reasonable prices. Sarajevo, the Bosnian capital that was surrounded by Serbian forces, may have reminded historians of the catastrophe of 1914, but most people remembered it as the site of the Winter Olympics of 1984. The scores of journalists who spent time in besieged Sarajevo reported that it was a cosmopolitan European city whose cultivated citizens suddenly found themselves trapped in a nightmare world that was supposed to have been gone forever.

The Balkan Wars of the 1990s had terrible consequences for those living in the region. But the wars' significance for the rest of Europe lies more in what did not happen than what did. The conflicts in the Balkans did not spread. In contrast to the years before 1914, the Balkans were not a conduit through which violence could move from the periphery to the center of the European society of states. A new era of armed conflict did not begin — at least not for Europeans.

But while the Balkan Wars did not do what some feared, they were also not the occasion for a vigorous and united European foreign policy. Instead, Europeans' fitful and ineffective interventions demonstrated that they were unable or unwilling to restore peace to the increasingly tormented peoples living along their southeastern frontier.

For much of the world, the 1990s were a time of great change: old regimes collapsed, new states were created, the geopolitical structure of the Middle East was transformed. Given all this turbulence, one is struck by the slow growth and remarkable continuity of European institutions, especially the consolidation of the European Union and the survival of NATO. Like the famously silent dog in the Arthur Conan Doyle story "Silver Blaze," the absence of radical change in these institutions gives us a clue with which to understand the European condition in the early twenty-first century. As we shall see, the values and assumptions of Europe's civilian state were tested but not transformed by the dangers and opportunities of the new era that began in 1989.

• • •

The end of the Cold War seemed to threaten the foundations on which the European international order had been built. In October 1990, the German-German border, for forty years the potential flashpoint for an East-West military conflict as well as the symbol of the superpowers' commitment to the postwar status quo, disappeared from the political map; the new Germany, although firmly bound to the West, was now completely free from external limitations on its sovereignty. A month later, at a meeting of thirty-four states in Paris, the members of NATO and the Warsaw Pact signed an agreement that ended the Soviet Union's long-standing superiority in conventional military forces. And this, as it turned out, was just the beginning: soon the Warsaw Pact, then the Soviet Union itself, disappeared. The Russian Federation, formed from the wreckage of the Soviet regime, still had a formidable — and worrisome — arsenal of nuclear weapons, but the armies that once menaced western Europe were gone.

Given the dramatic shift in the distribution of power in Europe after 1989, it is not surprising that political analysts like John Mearsheimer, who believe that "the distribution and character of military power are the root causes of war and peace," expected a new system to emerge from the wreckage of the Cold War order. In 1994, Kenneth Waltz, another influential advocate of the "realist" approach to international relations, wrote that the world would now have to learn to do without the "stark simplicities and comforting symmetry" of the old bipolar division between East and West. Without the stabilizing authority of the two superpowers, Waltz argued, the leaders of the European states will "relearn their old great power roles." When this happens, the inherent anarchy of the international system will return, shattering the institutional framework that had been created in the late forties and early fifties. "The prospects for major crises and war in Europe," Mearsheimer predicted, "are likely to increase markedly."

Contrary to these expectations, the institutional integration of postwar Europe not only survived the end of the Cold War but became stronger and more extensive. Even the newly unified Germany — indeed, especially Germany — which the realists regarded as the state most likely to reclaim great-power status, was eager to remain in the European framework that had served it so well. As Richard von Weizsäcker, the president of the German Federal Republic, declared in October 1990, "Today sovereignty means participating in the international community." Instead of relearning their old roles, European statesmen continued to perform the parts they had mastered during the preceding four decades; these were, after all, the parts their electorates expected them to play and the ones that had produced unprecedented peace, prosperity, and stability.

In July 1990, leaders of the NATO governments gathered in London to ratify a "New Strategic Concept" designed to meet "the profound political changes" that had "radically improved" the alliance's security environment. This "Concept" was expanded and absorbed into a "Declaration on Peace and Cooperation," issued after a summit meeting held in Rome on November 7 and 8, 1991. While these two documents affirmed the continuing significance of the Atlantic alliance, they also acknowledged the need to reconsider its mission. In the first place, the new security environment would enable the member nations to cut the size of their armed forces, emphasize speed and flexibility instead of a forward defense, and reduce their dependence on nuclear weapons. Second, while the members maintained that "the military dimension remains essential," they clearly shifted their focus to the organization's political goals. These included dialogue with NATO's former antagonists in the East as well as greater cooperation with other European organizations. Behind these changes was a subtle but definite shift in the meaning of international security, which had increasingly become a problem of maintaining order and stability rather than defending territory against aggression.

With the end of Cold War, therefore, Europe had become a less dangerous but also a more complicated place, with alternatives and opportunities that had not been present when hundreds of thousands of Soviet troops were deployed along the East-West frontier. Foremost among these opportunities was the chance to revise Europe's always unequal and often contentious relationship with the United States.

Since its formation in 1949, the Atlantic alliance had faced one crisis after another. Washington and its European allies had disagreed about German rearmament and French defection, the invasion of Suez and the war in Vietnam, Kennedy's missile crisis, Nixon's détente, and Reagan's Strategic Defense Initiative. Again and again, Europeans weighed the costs and benefits of their dependence on the United States, calculating and recalculating the balance between the security provided by American power and the dangers American policies frequently imposed. But as much as some Europeans would have liked to bid their partner across the sea a fond farewell, so long as the Soviet threat remained, few had a convincing answer to the question posed in the title of a book by the German politician Walther Kiep in 1972: *Good-bye Amerika, was dann?*—"Goodbye America, then what?"

Many Europeans believed that the end of the Cold War made it possible to find new answers to Kiep's question. A united Europe could become powerful enough to be an equal partner with the United States and, if necessary, to play an autonomous role on the international stage. Efforts to create a new set of European institutions were already under way before 1989. In 1986, members of the European Economic Community signed the Single Europe Act, which represented the most dramatic move toward integration since the Treaty of Rome almost thirty years earlier. The act laid out an ambitious program for removing the remaining economic barriers within the Community and called for a number of reforms in the way policies were formulated and administered. Its ultimate goal was a European economy in which goods, capital, and labor moved without restraint *and* a political community in which major questions of domestic and foreign affairs would be addressed by Europewide institutions.

The momentum toward greater integration was not broken by the fall of communism. In December 1991, the aspirations articulated in the Single Europe Act were extended when the European Union was established by a treaty that is conventionally named after the Dutch town of Maastricht, where it was signed. This treaty was slightly modified and the Union's various foundational documents were greatly clarified by the Treaty of Amsterdam, which was accepted by the states in October 1997 and went into effect two years later. These agreements emphasized the importance of an enhanced European role in foreign affairs. The Single Europe Act, for instance, committed the Community to attempting "to formulate and implement a European foreign policy." And the Maastricht Treaty announced the Union's intention to construct "a common foreign and security policy including the eventual framing of a common defense policy, which might in time lead to a common defense, thereby reinforcing European identity and its independence [in order] to promote peace, security, and progress in Europe and the world."

The tentative quality of this pronouncement ("might in time") reflected Europeans' divisions about the nature of the new policy and especially about its implications for NATO. After considerable debate, the European Union's leaders agreed to shift responsibility "to elaborate and implement" security policy to the Western European Union, an organization that had been formed in 1948 but largely moribund since. The WEU included all the European states that belonged to the European Union and to NATO and thus seemed to provide a neutral forum in which security issues could be discussed. As quickly became apparent, the WEU's sudden revival was a symptom of, rather than a solution to, the difficulties raised by Europe's aspirations for identity and independence.

In November 1991, NATO's "Declaration on Peace and Cooperation" had officially welcomed European defense initiatives, including the expanded role of the Western European Union. The New Strategic Concept promised to "facilitate the necessary complementarity between the Alliance and the emerging defense component of the European integration process." Since "complementarity" can mean many things, the term veiled the deep disagreements among the Europeans and between them and their American allies. Washington would have been delighted to see Europeans accept more responsibility for their own security, but, as the George H. W. Bush administration warned its partners in February 1991, the Americans opposed any independent European security agency that would compete with NATO. For their part, the Europeans were divided between those, like the British and the Dutch, who continued to regard NATO as the primary source of their security, and the French and the Germans, who were prepared to move toward a more autonomous European security system. All the states, on both sides of the Atlantic, said that they wanted greater European independence, but deep differences remained about its meaning and implications.

The initial debates about European security after the Cold War took place in the midst of two international crises. The first was triggered by Iraq's invasion of Kuwait in August 1990. By annexing this small but oil-rich kingdom, Saddam Hussein, the leader of Iraq's repressive regime, hoped to acquire the resources necessary to recover from the costly but fruitless war he had waged against Iran between 1980 and 1988. The occupation of Kuwait put Saddam in control of 20 percent of the world's oil reserves and in a position to threaten another 30 percent in neighboring Saudi Arabia. From the start, Europeans—with the vigorous exception of Prime Minister Margaret Thatcher—were content to leave the initiative in the hands of the United States, which constructed a broad international coalition to support military action against Iraq. Some European states, including France and Britain, sent troops as part of this joint effort, but neither NATO nor the institutions of the European Union were involved. The coalition's swift and decisive victory in January 1991, therefore, underscored both American power and European dependence.

Unlike the war against Iraq, the conflict in the Balkans, the second major crisis of the 1990s, was too close for Europeans to ignore and could not easily be outsourced to Washington. In 1990 and 1991, the United States, which was still absorbed with the liquidation of the Cold War in Europe and the ongoing crisis in the Persian Gulf, let the Europeans know that Yugoslavia was *their* problem. With a pride the ancient Greeks would have recognized, the Europeans eagerly accepted the challenge. Jacques Poos, Luxembourg's foreign minister, declared, "The hour of Europe has dawned." In fact, the Europeans' interventions in the Balkans did little more than reveal their own uncertainties and divisions. Their most decisive action, taken at the insistence of Germany, was to recognize the secession of Slovenia and Croatia from Yugoslavia at the end of 1991. By helping to dismantle Yugoslavia's federal structure without considering what might take its place, the diplomatic recognition of the two breakaway republics encouraged competing factions to seize territory and expel other ethnic groups. At the same time, it intensified the predatory war being fought by Serbs and Croatians against Bosnia. As the former Yugoslavia's long agony unfolded, various agencies—the European Union, the Western European Union, the United Nations, and a number of ad hoc groups—tried to end the bloodshed with a bewildering series of plans, which merely enabled the warring parties to play one set of peacemakers against another.

In the face of that complex and mercurial situation, the European Union's institutions proved too cumbersome to formulate a unified, coherent policy. But more important than the Union's institutional weaknesses were the Europeans' failures of will and imagination. Their leaders seemed to lack the will to do more than talk, threaten, and condemn. When some European governments sent troops as part of the United Nations peacekeeping mission, they imposed restrictive rules of engagement that rendered the soldiers virtually helpless. Because they viewed the situation from the perspective of their own civilian states, the Europeans could not imagine what was at stake in the Balkans, nor could they understand the vicious but nonetheless rational calculations that drove people like Tudjman and Milosević to embrace violent solutions. Instead, Europe's leaders explained the Balkan Wars in terms of ancient ethnic hatreds, clashing civilizations, or primitive propensities to violence, which effectively absolved them of responsibility. After all, against such bred-in-the-bone cultural traits, what could civilized people be expected to do? "The conflict in Bosnia," Prime Minister John Major of Britain decided, "was a product of impersonal and inevitable forces beyond anyone's control." Major's comment reminds us of Lloyd George on the origins of the First World War. Fatalism is often a mask for failure.

In 1994, after more than two years of bloodshed, the situation in the Balkans began to change. That February, the United States government reluctantly started to pay attention to the increasingly desperate condition of the Bosnian Muslims. Following Washington's lead, NATO forces threatened to bomb Serbian positions if the Serbs did not cease their shelling of Sarajevo. Soon thereafter, the Bosnian government reached an agreement with Croatia, whose greatly improved armed forces had begun to shift the military balance. Meanwhile, the Clinton administration, under growing public pressure to do something to stop the killing, used a combination of promises and threats to separate Milosović from his ethnic Serbian allies in Bosnia.

The turning point came in July 1995, when the Bosnian town of Srebenica, designated by the United Nations as a "safe haven," was captured by Serbian soldiers, who brushed aside the Dutch troops sent to maintain the peace and then slaughtered thousands of helpless Muslim men and boys. This atrocity convinced American policy makers that the cost of inaction was too high. When Bosnian Serb forces ignored demands for a ceasefire, Washington ordered air strikes by NATO planes. Abandoned by their patrons in Belgrade, threatened by Croatian and Bosnian infantry, and under assault by the world's most powerful air force, the Bosnian Serbs finally agreed to stop fighting. In November, the various sides met at an American air force base outside Dayton, Ohio, and, after long and arduous negotiations, signed a treaty establishing Bosnian independence, based on a territorial division according to ethnicity. NATO contributed sixty thousand troops to a multinational force that guaranteed the settlement.

The resolution of the Bosnian conflict demonstrated the continuing significance of NATO for European security. At the end of 1995, the French government announced that it would resume cooperating with the alliance in ways that "do not encroach on France's sovereignty." But France had not abandoned its aspirations to make Europe less dependent on the United States. In an article published in the *NATO Review* in May 1996, the French minister of defense stressed that his nation's renewed role in NATO made it all the more important to clarify the alliance's European dimension. "The political scheme for European integration necessarily means that security and defense matters will increasingly be dealt with at European Union level," a development that was itself, he argued, "a means of consolidating the alliance and placing transatlantic solidarity on a sounder footing." The following month, a meeting of NATO defense ministers promised to create "a European Security and Defense Identity within the Alliance." In an effort to satisfy both the Europeans' demands for "identity" and the American insistence that this identity remain "within" the alliance, the ministers agreed to design command and support structures that would be "separable but not separate," thereby enabling independent European operations without duplicating or diminishing NATO.

The strains within the alliance, however, remained perceptible behind the diplomatic language of the policy statements issued by the European Union and NATO over the next three years. In 1997, the Treaty of Amsterdam listed among the Union's objectives (mentioned just after "promoting economic and social progress") the assertion of its identity on the international scene, including "the progressive framing of a common defense policy, which might lead to a common defense." Meeting at St. Malo in May 1998, President Jacques Chirac and Prime Minister Tony Blair removed the uncertainty suggested in "might lead to" by agreeing that "the Union must have the capacity for autonomous action, backed up by credible military forces, the means to decide to use them and a readiness to do so." Considering that Britain had always been a staunch supporter of NATO's Atlantic dimension, Blair's step seemed to mark a major move toward the building of an autonomous security capability.

When the leaders of NATO met in Washington to commemorate the alliance's fiftieth anniversary in April 1999, they issued yet another New Strategic Concept, which once again acknowledged European aspirations but repeated that these aspirations must be realized within the alliance's existing organization. On a more practical level, NATO made a few adjustments so that its communication network and other facilities might be deployed more easily by a European Union force. These concessions were clearly not enough for those who wanted greater European autonomy. At the EU summit held that June in Cologne, the European heads of state reaffirmed their commitment to having "a capacity for autonomous action supported by credible military forces," which meant being able to make decisions and to act "without prejudice to actions taken by NATO." Not unreasonably, American policy makers regarded this announcement as a repudiation of the agreements made a few weeks earlier in Washington.

Just as the initial post-1989 debates on European security were enmeshed with the first Gulf War and the dissolution of Yugoslavia, the debates at the end of the decade took place in the context of a new crisis in the Balkans. This time the site was Kosovo, a province with an Albanian Muslim majority that was legally part of the Serb-dominated Yugoslav Federation and historically the center of Serbian national identity. During the wars of Yugoslav secession, relations had deteriorated between Kosovo's privileged Serbian minority and its Albanian majority. At first, the Kosovar civil rights movement was nonviolent, but after 1995, a cycle of violent protests and still more violent repression began.

Repeated attempts at international mediation failed. Once again,

Victims. Kosovar refugees, 1999.

television screens around the world carried pictures of burned villages, broken bodies, and long lines of frightened refugees, some of them wearing traditional garb, others in warm-up suits and brightly colored anoraks. By 1998, after a quarter of a million Kosovars had been driven across the border into Albania, public pressure for intervention mounted in the United States and Europe. Following the collapse of peace talks at the beginning of 1999, NATO began to bomb targets in Serbia. What was supposed to be a brief exercise in coercive diplomacy turned into a seventy-eight-day air war, at the end of which Milosević finally agreed to remove Serbian forces from the province. The war was terribly costly, both for the Kosovars, who were subjected to the fury of Serbian troops throughout the campaign, and for the hundreds of thousands of Serbian civilians who had to endure the bombing. NATO forces, however, did not suffer a single battle fatality.

Europeans and Americans drew different lessons from the first war in NATO's history. In Washington, military strategists were encouraged by their ability to wage war without casualties but were unsettled, too, by the structural strains in the alliance. As we have seen, the commander of NATO forces was also the commander of American troops in Europe—a fusion of roles that had been the strongest expression of the United States' commitment to the continent's defense. In 1999 this post was held by Wesley Clark, a smart, ambitious, and hard-driving officer who was eager to end his brilliant military career with a decisive victory over the Serbs. When the bombing campaign dragged on longer than anyone anticipated, Clark found himself in an increasingly difficult position: he disagreed with his superiors in Washington, who wanted to avoid American casualties at almost any price; he distrusted his European allies, with whom he did not fully share intelligence; and after hostilities had ended, he was defied by the commander of the British contingent, who refused to obey his order to prevent the landing of Russian troops at Kosovo's main airport. In the end, these structural flaws did not matter because NATO won the war so easily. But in a more challenging military operation, the inherent problems in the NATO command structure could easily have had devastating consequences.

The war in Kosovo once again demonstrated Europeans' inability to act alone and thus deepened their conviction that they needed their own security system. At the European Union summit in Helsinki, held in December 1999, the participants agreed to establish a multinational force of around fifty thousand that could be mobilized in two months and sustained for a year. Three groups were set up to handle policy-making: an Interim Political and Security Committee at the ambassadorial level, an Interim Military Committee, and a Military Staff to provide advice on possible European operations. Relations between the EU security institutions and NATO were still obscured by clouds of imprecise rhetoric, but Europeans seemed ready to act more autonomously and independently.

But while the war in Kosovo may have strengthened Europeans' aspirations for greater independence, it also revealed the considerable gap between their power and that of the United States. The global role of the Americans, declared the French foreign minister, Hubert Vedrine, in 1999, "is not comparable, in terms of power and influence, to anything known in human history." Vedrine was certainly correct. At the end of the twentieth century, the United States stood alone, its rival superpower defeated and in disarray, its political, economic, and cultural authority apparent everywhere. U.S. military power was especially impressive: there were American bases on every continent, American warships sailed every ocean. In the nineties, American soldiers engaged in more than one hundred operations in fifty countries, five times more than in the last decade of the Cold War. Moreover, the United States' mastery of the new technology of warfare, called the "Revolution in Military Affairs," not only gave its forces an overwhelming advantage on the battlefield, but seemed to make it possible to inflict extraordinary damage on the enemy without suffering significant casualties. Victories in Iraq, Bosnia, and Kosovo convinced many experts that the American way of war, which had always favored the use of machines in order to preserve the lives of its citizen soldiers, had reached a new level of effectiveness. Europeans were still trying to come to terms with the strategic and political implications of this preponderance of American power when, like everything else in the international society of states, the Atlantic alliance was shaken to the core by the events of September 11, 2001.

Initially, Europeans responded with sympathy and support to the terrorist attacks on the World Trade Center in New York and the Pentagon in Washington. In two long meetings, the permanent representatives of NATO agreed, for the first time in the alliance's history, to invoke the mutual defense clause in Article 5. NATO warplanes were sent to patrol American air space. With a few marginal exceptions, the European press, public, and political elite expressed outrage at the attacks and solidarity with the United States. President Chirac and his cabinet ministers attended services at the American Church in Paris; pro-American banners were draped across Berlin's Brandenburg Gate. A month after the attacks, in a major address to the German parliament, Chancellor Gerhard Schröder expressed his nation's "unrestricted solidarity" and promised military assistance to defend freedom and security. Almost two thirds of the German public approved of the chancellor's offer of support, including his willingness to deploy German troops in Afghanistan.

Nevertheless, behind these proclamations of support lurked differences in the way Americans and Europeans viewed terrorism and how to combat it. Americans tended to see terrorism as a global movement that directly threatened their national security. To defeat it would require a war like the one that had destroyed the Axis powers in the Second World War—a comparison underscored by the constant association of September 11 with the Japanese bombing of Pearl Harbor. Europeans, who had been fighting their own local forms of terrorism for several decades, were inclined to see it as a persistent challenge to domestic order rather than an immediate international threat. The proper remedy was more effective policing, stricter laws, better surveillance. They wanted to extradite terrorists and try them as criminals, not wage war against states that were suspected of supporting them. The notion of a "war on terrorism" was misleading, warned Michael Howard, a British military historian, because the word "war" "arouses an expectation and a demand for military action against some easily identifiable adversary . . . leading to decisive results." Few Europeans doubted that terrorism was a serious issue, but most did not accept the official American position that a global struggle for national survival had begun on September 11.

Despite NATO's support for the United States immediately following the attacks on New York and Washington, only the British played a prominent part in the first military operation against global terrorism, the campaign to remove the radical Islamist Taliban regime in Afghanistan, which had provided a safe haven for Osama bin Laden, the architect of the September 11 attacks. Remembering the practical problems and political tensions that attended the Kosovo campaign, the Americans did not want to be burdened with allies in the kind of war they expected to fight, a war with a relatively small number of elite troops, lots of precision bombing, and a diverse array of local partners. On September 27, Deputy Secretary of Defense Paul Wolfowitz told NATO defense ministers that the alliance's support was not needed in Afghanistan. The war on terror, he added, "would be made up of many different coalitions in different parts of the world." When the Taliban were driven from power in a remarkably short time, it appeared that, once again, the United States had the capacity to win quick, low-cost victories without burdening itself with the Atlantic alliance's tangled chain of command and legally freighted rules of engagement.

The war on terror gave new urgency to the question of whether NATO had outlived its usefulness. Could an alliance that had been designed to counter a massive Soviet invasion of Europe be effective in a protracted struggle against rogue states, elusive bands of insurgents, and individual terrorists? Finding an answer to this question was the main purpose of the summit held in Prague in late November 2002, the first meeting of NATO heads of state since 1999 and one of the most important in the organization's long history. Prague, said an American official before the meeting, is about "deconstructing the old rigid NATO and building a new one, able to fight the wars of the future." A French diplomat made the same point when he remarked that the alliance had to show it was "not a creaky old lady, but a fresh young maiden, and that requires a lot of plastic surgery, with all the risks and pain."

With the excessive optimism and self-confidence characteristic of such documents, the final communiqué of the Prague summit committed the alliance to "transforming NATO with new members, new capabilities and new relationships with our partners." The new members were Bulgaria, Estonia, Latvia, Lithuania, Romania, Slovakia, and Slovenia, which brought the total number of states in the alliance to twenty-six. New capabilities included a streamlined command structure, a Response Force of combined air, land, and sea units that could move quickly to wherever it was needed, and the adoption of more technologically sophisticated weapons and support systems. New relationships involved working with former antagonists like Russia, as well as enhancing cooperation with the European Union and other international bodies. Altogether, the long and unusually detailed declaration represented NATO's determination "to meet the grave new threats and profound security challenges of the 21st century."

One word was conspicuously absent from the Prague communiqué: Iraq. No mention was made of the issue that, by November 2002, had begun to dominate Atlantic relations and would eventually come close to destroying the alliance. Ever since the end of the first Gulf War in 1991, there had been serious disagreements between Washington and its allies about the proper policy toward Saddam Hussein. The United States, usually backed by Britain, was in favor of a stern policy of containment, including economic sanctions, the imposition of a strict no-fly zone to protect Iraq's Kurdish minority, and occasional aerial bombing to punish Saddam's failure to cooperate with U.N.-mandated weapons inspections. Some American policy makers did not think containment was enough and were eager to destroy Saddam's regime. Many European governments, especially the French and the Germans, not only believed that Saddam could be contained, but were prepared to seek Iraq's gradual reengagement with the international community, even if this meant easing sanctions and halting punitive air strikes.

After September 11, these disagreements were intensified by Washington's decision to make Iraq a prime target in the war on terror and to use "regime change" in Baghdad as a way of transforming the political constellation in the Middle East. But the issue dividing the alliance was never merely Iraq. Just as Vietnam had once stood for an ensemble of things people feared and distrusted about the United States, so Iraq symbolized a number of American policies with which Europeans disagreed: an excessive reliance on military solutions, the threat of preemptive action, and the apparent disregard for consultation and cooperation. Added to these were the death penalty, the absence of gun control, inattention to global pollution, and a growing collection of other irritants, great and small, real and imagined, with which America had become identified.

Divisions among European governments, however, were at least as deep as those between Europe and the United States. As usual, the British were closest to Washington and, alone among the NATO powers, had the capacity to provide significant military assistance. France, predictably, was most eager to develop an independent policy. President Chirac, like François Mitterrand in 1991, took the lead in seeking a diplomatic alternative to war; but unlike his predecessor, Chirac showed no signs of being willing to participate in military operations should diplomacy fail. Chancellor Schröder, engaged in a hard-fought battle for reelection and confronted with serious economic problems at home, wooed voters by campaigning vigorously against America's emphasis on military solutions, in Iraq and elsewhere. Relations between Washington and Berlin reached an all-time low when a member of Schröder's cabinet compared President Bush to Hitler.

In November 2002, just a few days before the Prague summit, Washington and its European allies achieved a fragile consensus based on U.N. Security Council Resolution 1441, which charged Saddam

—erroneously, as it turned out—with violating earlier resolutions by not surrendering his weapons of mass destruction, and gave him the opportunity to participate in a new, strictly defined inspections regime. Opponents of military action supported the resolution because it seemed to offer Saddam the chance to avoid war. The Americans and the British, who had already decided that war was the only option, used the resolution to secure the legal and political basis for armed intervention.

By January 2003, when the massive deployment of American forces in the Gulf had made war inescapable, the divisions in the alliance became more pronounced and more public. At a Paris press conference, Chirac and Schröder condemned American policy; a few days later, eight European heads of state signed a declaration in support of Washington. At the end of the month, when the United States asked its NATO allies to take some limited steps in support of the impending operation, France, Germany, and Belgium refused, arguing that even planning implied approval of the war. Eventually a tentative compromise was reached, but what the American ambassador to NATO called a "crisis of credibility" in the alliance remained. The conflict over Iraq also produced a quieter but no less serious crisis of credibility in the European Union: as the contending sides coordinated and articulated their positions, no one paid attention to the institutions that were supposed to create a common European foreign policy.

Only three European governments—France, Germany, and Belgium—actively opposed the war in Iraq; the rest responded with varying degrees of support or at least compliance. But the overwhelming majority of Europeans, including those whose governments had joined the American-led coalition, were strongly and often vocally against military action. Opinion polls in nearly every European state showed the same trajectory: in Germany, 78 percent of the public had a positive view of the United States in 2000; by 2002 this had dropped to 61 percent; by the time the war had started, to 25 percent. In Spain, whose government belonged to the "coalition of the willing," 14 percent of the population viewed the United States favorably. In Turkey, which had refused to allow American troops to be deployed on its territory, 12 percent of the public supported American policy.

At first, the war in Iraq, which began in March 2003, seemed as if it would be another quick and relatively cheap victory for American military technology. Within a few weeks, however, it became apparent that defeating the Iraqi army was only the beginning. Iraq became engulfed by insurgency, in which diehard defenders of Saddam's regime, Shia and Sunni militias, and a variety of foreign volunteers fought occupation forces and one another. The Kurdish areas in the north, which had already achieved a degree of autonomy following the first Gulf War, were largely stable and orderly, but in many other areas occupation troops and their Iraqi allies were under constant attack. Although American officials talked a good deal about the international makeup of the coalition, most of the fighting was carried on by American and, to a lesser degree, British soldiers. As the military and political situation deteriorated, the other members of the coalition either reduced or withdrew their small contingents.

The prolonged agony of this second Gulf War increased public hostility to the United States in Europe (and elsewhere), but also made officials in Washington more aware of the advantages of international cooperation. Perhaps the most important result of this was the engagement of NATO forces in Afghanistan, where, following the deceptively swift defeat of the Taliban regime, the new government still required foreign assistance to maintain order and hold on to power. Although NATO took responsibility for operations in Afghanistan in August 2003, American troops remained deployed in the south, where most of the fighting was taking place. In May 2006, a British general assumed command of the NATO forces, and more and more European units moved into violently contested areas. The outcome is as yet unclear. But it is clear that this is not an air war like the Kosovo conflict, and not a peacekeeping mission like Bosnia. For the first time in its history, NATO is fighting a ground war in which substantial battlefield losses are possible. A failure to meet these challenges, either militarily or politically, would put NATO's future at risk. Afghanistan, in whose harsh terrain so many dreams of empire have been interred, might then become the graveyard for the Atlantic alliance.

In an article published in the spring of 2003, at the height of the Iraq crisis, the British political scientist Stephen Haseler expressed an increasingly widespread conviction among Europeans: "Europe does not need America—any more than America needs Europe—for its fundamental defense and security." This has been true, Haseler argued, since 1989, but as a result of the second Iraq War, it has become the basis for a new European foreign policy. The first concrete expression of this policy was the Franco-German condominium against the war; Haseler predicted this anti-American alliance would eventually spread across Europe until it even included Britain, Washington's most loyal European partner. Within five years, he suggested, Europe would have a new security system, with its own command structure, Rapid Reaction Force, nuclear strategy, communications satellite, airlift capability, and intelligence. "Such a security system," he added, "cannot proceed for too long simply on the basis of inter-governmental co-operation. Those EU members who seek a common security and foreign policy system will need, sooner or later, to move towards a supranational structure." Haseler's final point is worth emphasizing: because of the high costs of military preparedness, strategic independence from the United States cannot be achieved by individual states; European autonomy requires greater European integration. No one state, even a rich and potentially powerful one like Germany, can do it alone.

In 2003 there were indications that Europeans were willing to project their collective power beyond the Union's borders. In March, *European* troops went into action for the first time in history when 320 soldiers, wearing EU insignias on their national uniforms, took over peacekeeping duties in Macedonia. That June, another European force carried out a three-month mission in the Democratic Republic of the Congo. At the end of the year, a European summit in Brussels endorsed a statement prepared by Javier Solana, the Union's chief foreign policy official, which included a commitment "to develop a strategic culture that fosters early, rapid, and when necessary, robust intervention." More impressive than the proclamation of these ambitious goals—which were, after all, a familiar feature of all European statements about security—was Europe's assumption of peacekeeping duties in Bosnia with a force of seven thousand. "Bosnia," said Geoff Hoon, Britain's defense secretary, "is a demonstration in practice that this [EU defense policy] can work."

Although Europeans made some progress toward the development of an independent security system in 2003 and 2004, their accomplishments still lagged far behind the frequently voiced aspirations of their leaders. The operations in the Balkans were small, the environment relatively benign, and the political decisions to deploy rather straightforward. The missions in Macedonia and Bosnia did not demonstrate that the European Union would be able to intervene in a major crisis. The Rapid Reaction Force that was supposed to be ready by the end of 2003 was still not operational, and in a number of critical areas—airlift capacity, communications, intelligence gathering, precision bombing—the Europeans remained woefully ill prepared. In their training, equipment, and operations, European forces were more like heavily armed police officers than military units.

What would it take for the Europeans to create a genuinely independent security system?

Clearly, this system could not be based on the mass conscript army that had dominated strategic theory and practice throughout most of the twentieth century. As we have seen, conscript armies had been withering away well before the end of the Cold War. In the 1990s, as governments recognized that their armies were neither financially nor militarily efficient, more and more of them either abolished the draft or reduced the period of active service to a few months. The European model is now the kind of relatively small, professional force that the British developed after they abolished the draft in 1960. But to be effective, this force must have the computers, laser-guided weapons, and communications capabilities that a technologically driven war now requires.

This military technology is very expensive. The Rapid Reaction Force, the rough equivalent of a U.S. Marine brigade, would cost upward of $50 billion. There is little evidence that most European states are willing to spend that kind of money. In fact, while the cost of military hardware continues to expand, European military budgets keep shrinking. Between 1985 and 1999, France's defense budget declined 7 percent, Germany's 15 percent, and Britain's 40 percent. In 2003, when the United States spent about 3.3 percent of its GDP on the military, Germany spent 1.5 percent. At this level of spending, most of the money is necessarily absorbed by fixed personnel costs, which in some European states account for up to 70 percent of the total defense budget. Germany, for example, spends almost two thirds of its budget on personnel, including 130,000 civilian employees, who work 220 days a year and have guaranteed jobs for life. Only about 13 percent of the German budget is spent on new equipment, much less on research and development. Not surprisingly, the Germans have a limited number of combat-ready troops and no capacity to deploy them. In order to send peacekeepers to Afghanistan, they had to lease aircraft from Ukraine. In Germany, as well as in the rest of the European Union, either substantial new expenditures or a painful reallocation of resources would be necessary to reverse these trends.

The European Union is the largest economic bloc in the world; its members create about one quarter of the world's gross national product and one fifth of the world's commerce. With a large and productive population of 445 million and a combined GDP of about $11 trillion, there is no doubt that the members of the Union can afford to fund an effective military force. There is considerable doubt, however, about their willingness to devote a larger share of their resources to defense, particularly considering the growing fiscal constraints confronting every state. While an overwhelming majority of Europeans say they want Europe to be a superpower, only a third of them want to spend more money on defense. Meanwhile, the military's share of government expenditures continues to decline. In Germany, which already has the lowest per capita military spending among the major European states, the 2004 defense budget was reduced by more than $3 billion from the year before.

Europeans could, with some difficulty, find the resources necessary to improve their military capacities if they could overcome the even harder challenge of creating effective decision-making institutions. To get a sense of this challenge, one has only to look at the way foreign and security policy was treated in the European constitution that was drafted in 2003, debated and passed at a Union summit in 2004, and then rejected by referendums in France and the Netherlands in June 2005. The constitution would have established a Union minister of foreign affairs, who was to be appointed by the European Council and would also serve as a vice president of the European Commission; together with the Commission as a whole, the minister would be subject to the approval of the European parliament. Like the comparable national official, the minister would

be in charge of the Union's foreign policy, but the policy would apparently—the constitution is not entirely clear on this point—need
the unanimous support of the European Council, much as the U.N.
secretary-general needs the unanimous support of the permanent
members of the Security Council. The constitution's discussion of
the possible use of force is even more cautious. No state could have
been compelled to go to war by a majority vote of the council, although the promise of a "progressive framing of defense" left open
the possibility of more cohesive military institutions.

The constitution's sometimes willfully obscure and sometimes
sharply qualified definition of how a common foreign policy would
be made reveals the European states' reluctance to surrender their
autonomy in this key area. Even if the constitution had been accepted, it seems unlikely that its cumbersome mechanisms would
have been able to absorb the various interests, traditions, and institutional cultures that influence policy-making by its members.
The Union's experiences in the Yugoslav and Iraq crises do not offer
much hope that, in the midst of the unstable, potentially violent
world of international politics, it could impose a common policy on
its disparate members, keeping them together and on course.

Just as making decisions about security policy puts a greater burden on institutions than, say, making decisions about fiscal or agricultural policy, justifying these decisions requires a much higher level
of legitimacy. The tensile strength of a system's legitimacy is measured by the political weight it must carry, and nothing is heavier
than the question of peace and war.

The problem of legitimacy brings us to the "democracy deficit"
that is usually regarded as the European project's most conspicu-
ous weakness. The European parliament, based in Strasbourg, has
always been the Community's least effective institution. Originally,
it was indirectly elected by the legislatures of the individual states;
even after direct elections were introduced in 1979, there has not
been a unified system of voting, clear-cut party alignments, or well-
defined European issues. In contrast to the European Court, which
has operated quietly to weave an integrating fabric of rules across
the Community, the parliament's proceedings are public but largely
ineffectual. Its manifest limitations are a constant reminder that
the new Europe is not a representative democracy because it lacks

a demos to represent, a body of engaged citizens whose primary political loyalty is to Europe.

Even more than the citizenship of individual states, membership
in the EU is a matter of rights and privileges, not obligations and
commitments. At the beginning of the twentieth century, people's
political identities were shaped by habits, rituals, and institutions
designed to reinforce loyalty and commitment to a particular state.
One learned to be French or German or Italian in school, in the
army, and in the streets of every city, where monuments and civic
ceremonies taught the same patriotic lesson. Since 1945, such institutions and symbols have become progressively weaker in every
European state; in the European Union itself they have never existed. The Union makes no effort to forge an identity for its members. It does not demand that they be Europeans and not something
else. Instead, European identity is a diffuse amalgam of national, regional, and cultural allegiances in which no one element necessarily
predominates.

In the 1950s, at the very beginning of the integration process,
Raymond Aron wrote that "the European idea is empty, it has neither the transcendence of messianic ideologies nor the immanence
of concrete patriotism." Aron was half right. The idea of Europe did
not evoke emotional commitment. It did not stir people's hearts as
nations sometimes had done. It was not something for which many
would have been willing to give their lives. But the European idea
was not empty—or rather, it only seemed empty when compared to
the traditional idea of the nation. The European idea was full, not
of national enthusiasm and patriotic passion, but of a widespread
commitment to escape the destructive antagonisms of the past and
a deep concern for those economic interests and personal aspirations
that dominated public life in the second half of the twentieth century. Because the European Union does not claim Carl Schmitt's
"monstrous capacity," the power of life and death, it does not need
citizens who are prepared to kill and die. It needs only consumers
and producers, who recognize that the community serves their interests and advances their individual well-being. And as consumers
and producers, most Europeans have usually been rather satisfied
with the Union's accomplishments.

Viewed from the perspective of twentieth-century European history, we can understand why the European Union is not a super-power and why it is not likely to become one in the foreseeable future. In the first half of the century, European states, as we have seen, were made by and for war—the war for which states prepared before 1914 and the two great wars they fought between 1914 and 1945. In the century's second half, European states were made by and for peace. Contrary to what some have argued or expected, European states have not disappeared. Indeed, they are more stable and effective than ever before. But they have been transformed: they are now dominated by civilian institutions, focused on civilian goals. European states still have armed forces—just as garrison states had economies—but politically, symbolically, and economically, these military institutions are subordinated to the agencies that do what citizens regard as important: managing the currency, promoting economic growth, providing welfare, and protecting people from life's vicissitudes. The eclipse of the willingness and ability to use violence that was once so central to statehood has created a new kind of European state, firmly rooted in new forms of public and private identity. As a result, the European Union may become a superstate—a super *civilian* state—but not a superpower.

Epilogue

The Future of the Civilian State

I T I S "less and less logical," wrote the French strategic expert Pascal Boniface about the European Union in 2000, "for an economic, commercial, technological power . . . to remain a minor power on the strategic plane. It amounts to a historical incongruity." Boniface's assumption, shared by many of those who lament the discrepancy between Europe's economic strength and military weakness, is that the natural end of economic and legal integration is the creation of something like a nation-state with a federal structure, a unified foreign policy, and independent military power. Anything else is illogical or incongruous. Europe's unprecedented prosperity and stability, therefore, has always been shadowed by a pervasive sense that the Union is somehow incomplete, its full promise unrealized, its historical task unfulfilled.

But why should economic integration necessarily lead to political unity? This did not, as is sometimes supposed, happen in nineteenth-century Germany: the German Empire that was created in 1871 did not grow "logically" out of the German Customs Union, but rather was the product of two wars, a war against Austria that established Prussian hegemony in central Europe and a war against France that quickened the process of nation-building. Wars—a war of independence against Britain, a civil war to establish northern hegemony, and two world wars that greatly increased the power of the federal government—also played a crucial role in the creation of the United States, another federal state that is often cited as a model for

Europe's future. But the European Union, to repeat a central argument of the previous three chapters, is not the product of war, but of peace.

There is, moreover, nothing illogical or incongruent about the co-existence of economic strength and military weakness in contemporary Europe. In fact, the two fit together, each reinforcing the other. The flourishing of the European economy, and of the common institutions that bind Europeans together, produced what I have called civilian states. The power that these states project, both individually and as part of the European Union, emphasizes commerce, law, and culture—in other words, the activities and values at the core of their civilian identities. To change the role that Europe plays in the world would require not merely new institutional arrangements—that was the fundamental error of the failed constitutional process—but also a new political identity, which would have to be embedded in a different sort of civic culture and expressed in a different kind of state.

It seems highly unlikely that Europeans will change their civilian identities. Why should they? The blend of commitment and coercion that once motivated people to fight and die for their nations is gone forever. And so is the threat of aggression by other European states: the passengers in Bismarck's carriage are no longer armed. Europe has become a "non-war community," which means that its citizens live in what Edward Luttwak has called a "post-heroic age," sustained by private ambitions and individual desires. In the early twenty-first century, national defense is no longer the duty of each citizen; it is a matter for professionals who are paid to assume the risks and bear the burdens that come with their jobs. Like police officers and firefighters, professional soldiers deal with emergencies that threaten civilian life. These professionals are necessary, even admirable, but no one would suppose that they represent the ideal citizen or that such emergency services are somehow "schools of the nation." Defending the civilian state has become a job like any other.

In the light of Europe's history over the past hundred years, is it surprising that most Europeans do not want to pay the price of heroism, risk what they have built, sacrifice their personal accomplishments? Europeans know, in a way few Americans can, what war is really like. And while the ranks of those who directly experienced the world wars are thinning, the collective memory of those terrible times—the London Blitz, the occupation of France, the battle of Berlin, the agony of the Warsaw uprising, and the prolonged purgatory of the eastern front—remains. Set against these memories, indeed set against any other period in European history, the creation of the civilian state seems like something to be cherished, celebrated, and preserved.

Such civilian states require peace and order. Their future, therefore, depends on whether political violence, which was a central part of Europe's history for so long, could return to the European society of states. At the moment, it seems difficult to imagine a situation in which the states would fight one another. Even without its Cold War incubator, the process of peaceful integration has proved robust enough to flourish on its own. If violence does erupt in Europe, it will not come from inside but from outside, from the unstable and dangerous world in which Europeans must live their civilian lives.

There have always been those who hoped that the eclipse of violence would spread from Europe to the rest of the world. In 1958, for example, Ernest Haas, an early student of European integration, called Europe "a living laboratory" in which one could observe the necessary connection of economic unification, political cooperation, and international understanding. Many continue to believe in the exemplary character of the European project. In a voice amplified by his own domestic discontents, the American writer Jeremy Rifkin recently proclaimed that "Europe has become the new 'city upon a hill.' The world is looking to this grand new experiment in transnational governance, hoping it might provide some much needed guidance on where humanity ought to be heading in a globalizing world."

As we have seen, Europe did serve as a model for many states on its periphery. The successful transitions to democracy in Greece, Portugal, and Spain were all strongly influenced by European values, economic power, and political influence. The magnetic attraction of European achievements was also felt along the western marches of the Soviet Empire, especially during the empire's terminal crisis. Equally important, the European Union's extraordinary enlargement to include most of the former Communist states of eastern

Europe has served to consolidate both democratic institutions and civilian values. We should not lose sight of this peaceful expansion of Europe's "soft power," which has done so much to create a prosperous and stable environment on much of the continent.

And yet, since the end of the Cold War we have also become aware of the limits to the spread of Europe's civilian values and institutions. In those areas where international and domestic violence, or at least its strong possibility, remains at the center of politics, civilian states cannot grow and flourish. This is obviously the case where political borders are still contested by rival states, and even more painfully so in those regions where states are at war with their own citizens. In places like Lebanon and Iraq, Kashmir and the Gaza Strip, western Sudan and the Republic of the Congo, security trumps all else because violence is an ever present part of life. Here, where what Ole Wæver called "desecuritization" has not occurred, civilian values and institutions still take second place to the struggle for security and survival. Sovereignty remains a precious goal to be defended or acquired, not something to be surrendered or shared. Open borders, transnational legal institutions, and similar political experiments make little sense.

As the American political scientist Robert Keohane has written, "One of the most vexing questions in Europe today is where the frontier between the West European zone of peace and the Eurasian zone of conflict will be." It is clear enough who is on the wrong side of this frontier: the fragments of the former Yugoslavia — Bosnia, Kosovo, Macedonia — where foreign troops still maintain a fragile peace; the disorderly areas on Russia's Caucasian edge — Ingushetia, Chechnya, Dagestan, whose political future is still contested, and the province of Nagorno-Karabakh, an Armenian enclave struggling to survive in Azerbaijan. Many other states could fall either way. Slovenia is certainly a civilian state, but what about Croatia and Serbia? The Baltic republics seem securely civilian, but what about Belarus, Ukraine, or, for that matter, Russia itself?

Perhaps the most pressing and challenging boundary question facing contemporary Europe is posed by Turkey, now an eager aspirant for membership in the European Union. Turkey is a difficult case not, as is sometimes argued, because it is a Muslim state, but because it is obviously not a civilian state. Modern Turkey was, as we have seen, created by the army; its founder, first president, and omnipresent historical icon is Kemal Atatürk, the architect of military victory. Throughout the life of the republic, the army has played a powerful political role as what is sometimes called "the deep state." On several occasions, most recently in 1997, the army has intervened in politics to defend the Kemalist legacy of which it regards itself as the true custodian. Unlike the members of the European Union (with the qualified exception of Greece), Turkey has a mass conscript army with nearly universal military service. And sustained political violence is a part of Turkey's recent past, when an estimated thirty thousand of its citizens, most of them Kurds, died during a prolonged civil war. Turkey is a valuable ally and a critically important member of NATO. It has a great deal to offer the European Union. But it will not be an easy matter to absorb this kind of state into Europe's resolutely civilian politics and culture.

There is, at the moment, no direct military threat to Europe from its periphery. But Europeans are vulnerable to other sorts of incursions: pollution, disease, and criminality can all pass easily through the Union's porous boundaries. The bombings that took place in Madrid in March 2004 and London in July 2005 reminded Europeans that they could also be victims of radical Islamic terrorism. A failure to integrate their increasingly héterogeneous populations would not only inject violence into Europeans' domestic life but establish destabilizing alliances between internal dissent and international unrest.

Clearly, Europeans cannot cut themselves off from the outside world, where they must find markets for their goods, the resources to fuel their machines, and the labor to supplement their own aging populations. And of course Europeans are tied to a wider world by a sense of shared humanity, a responsibility to use their wealth and power to help their neighbors live better lives. Europe may be an island of peace and plenty, but, in the poet's words, it is not entire of itself—geographically, economically, and morally, it is a piece of the continent, a part of the main.

At the beginning of the twenty-first century, the most crucial question facing Europeans is how they will live in this dangerous, violent world. As tempting as it may be, Europe cannot withdraw and become, in Timothy Garton Ash's words, like "a gated community for the rich surrounded by poorer neighborhoods and terrible slums." At the end of the previous chapter, I suggested how hard it would be for Europeans to create a military force powerful enough to enable them to play a truly independent role in the world. So it seems likely that Europe will continue to depend on some version of the Atlantic partnership, with all of its attendant tensions and conflicts. Indeed, the same civilian values and institutions that make it necessary for Europeans to depend on the United States make the terms of this dependence difficult to manage. Nevertheless, as hard as it might be to live with the Atlantic alliance, living without it would be harder still.

Since the 1950s, Europeans have enjoyed a period of peace and prosperity unparalleled in their history. Never before have so many of them lived so well and so few died because of political violence. Dreams of perpetual peace, born in the Enlightenment and sustained through some of the most destructive decades in history, seem finally to be realized. Of course there are no resting places in human affairs, nowhere to hide from the insistent pressure of change. To sustain their remarkable accomplishments, Europeans must face a number of economic, political, cultural, and environmental challenges. Many of these challenges come from, or are influenced by, that long and ill-defined frontier that joins Europe to its neighbors. Along this frontier, where affluence and poverty, law and violence, peace and war, continually meet and uneasily coexist, the future of Europe's civilian states will be determined.

Notes

Bibliography

Index

Notes

page PROLOGUE

xiii *largest demonstration:* My account of the demonstrations is based on the following newspapers: the *Guardian,* February 17, 2003; the *Observer,* February 16, 2003; *Berliner Zeitung,* February 17, 2003.

xv *a new "European nation":* Garton Ash, *Free World,* p. 46.
"counterbalance the hegemonic unilateralism": Habermas, *Westen,* p. 45. The Habermas-Derrida declaration was written by Habermas and published, under a different title, in the *Frankfurter Allgemeine Zeitung* on May 31, 2003.
"on major strategic and international questions": Kagan, *Paradise,* pp. 3–4. For a critical analysis of Kagan's argument, see the essays in Lindberg. On Kagan's career, see Packer, pp. 17–24.

xvi *Marshall Fund a poll by the German: Economist,* June 5, 2004.
"the last truly sovereign": Garton Ash, *Free World,* p. 119.

xvii *When Mueller first presented:* Mueller, *Retreat;* Keegan, *War;* and Howard, *Invention.* See also Joas and Michael Mandelbaum's book and article. The most recent version of Mueller's argument is *Remnants.*

xviii *"The great powers of our time":* Bond, p. 27.

xix *"What good are the best social reforms":* Friedrich Naumann in 1895, quoted in Mayer, *Weber,* pp. 45–46.

1. "WITHOUT WAR, THERE WOULD BE NO STATE"

3 *"Without war":* Treitschke, vol. 1, p. 72. On Treitschke's lectures, see Dorpalen, p. 226f.
"Every state known to us": Treitschke, vol. 1, p. 72.

4 Krüger's painting is in the National Gallery in Berlin. There is a reproduction in *Nationalgalerie Berlin,* p. 218.
"tests of manly discipline": Vogel, p. 29.
military displays were also common: On the Habsburgs, see Stone and Rothenberg.

5 *"this grand school of patriotism":* Vogel, p. 98.

6   *"the image of the nation":* Vogel, p. 98.
    "must be personified": Walzer, p. 194.
7   *"States make war":* Tilly, *Coercion,* chapter 3.
8   There is a good summary of Prussia's military reforms in Craig, *Politics.*
    "They are an embarrassment": Craig, *Königgrätz,* p. 38.
9   *"mass" and "preparation":* Ropp, p. 196.
    On French military reforms, see Challener, Ralston, and Pedroncini, vol. 3.
10 On the Italian army, see Ceva.
11 On Russia's army reforms, see Wildman.
    Russians "have adopted a system": Kiernan, p. 153.
    Britain had introduced: On the British army, see Dunlop.
    "When we speak": Dunlop, p. 129.
12 *"the turn of compulsion":* Luvaas, p. 311.
    "We have become": Glenny, p. 219.
    On the size of national armies, see the data in Van Creveld, *Command,*
    pp. 148–49; Ferguson, *Pity,* pp. 93–94; and Stevenson, pp. 39–40.
13 Figures on military budgets are from Stevenson, pp. 2–6.
14 For the influence of military service and citizenship, see Gosewinkel.
    On the role of military service in the expansion of the state, see Weber,
    Peasants, and Frevert.
15 On European armies in the nineteenth century, see Gooch, *Armies.*
16 Marshal Lyautey quoted in Ropp, p. 199.
    "Discipline will be": Zeldin, vol. 2, pp. 881–82.
17 *"twin brother . . . both of them":* Jouvenal, pp. 10–11.
    "contrary to appearance": Janowitz, p. 186.
    "school room for the factory": Maude, p. 92.
    "the very cornerstone": Maude, p. 13.
18 *"the task of instilling":* Zeldin, vol. 2, pp. 881–82.
    "loyal subjects . . . must work like a healing": Kiernan, pp. 148–49.
    "The army is the great crucible": Gooch, *Army,* p. 118.
19 On Haldane's reforms, see Spiers.
    "the spirit of militarism": Spiers, p. 137.
    "most of the English middle class": MacKenzie, p. 255.
    "To make the citizen": Wilkinson, p. 191.
    "a genuinely national army": Treitschke, vol. 2, p. 357.
    On military service and masculinity, see Frevert.
20 *"It is not in giving life":* Beauvoir, p. 64.
    "state's monstrous capacity": Schmitt, *Begriff,* pp. 45–46.

2. PACIFISM AND MILITARISM

22 *"a happy overture":* Lepsius et al., vol. 15, pp. 142–43. The best account of
    the two Hague conferences is Dülffer.
    "This suggestion": Lepsius et al., vol. 15, pp. 147–48.
    Lord Gough quoted in Dülffer, p. 76.

24   *"in questions of a legal nature"*: Proceedings, vol. 1, p. 159.
26   *"the idea of disarmament"*: Rich, vol. 2, p. 603.
      "At the conclusion of": Dülffer, p. 137.
27   *"War appears to be"*: Howard, *Invention*, p. 1.
      "the most fortunate war": Silberner, pp. 72–73.
      "nothing could be": Semmel, p. 72.
      "drawing men together": Silberner, p. 61.
      "Free trade": Condliffe, p. 197.
28   *"frontier brawls"*: Bloch, *Future*, ix.
29   *"Bloch's war"*: Werner, p. 100.
      "the stoppage of military": Bloch, *Future*, p. 91.
30   *"new star"*: Suttner, p. vi.
31   *"In peace time"*: Dülffer, p. 302.
      On Angell's influence, see Marrin, p. 107f.
32   *"International politics are"*: Angell, p. 51.
33   *"those moral or material ends"*: Angell, p. v.
      "I am not a non-resister": Ceadel, p. 179.
      "It is more to the general interest": Angell, p. 149.
34   *"gradually forget the faith"*: Laity, p. 192.
      "I wish I could": Marrin, p. 69.
35   *"something worth having"*: Delbrück, p. 229.
36   *"from the single policeman"*: Mahan, pp. 31, 39.
      "we were living in . . . A nation too selfish": Chesney in Clark, pp. 47–48.
37   *"a roaring wave of fear"*: Wells, *War*, p. 91.
      "No man": Wells, *First and Last Things*, pp. 205–6.
38   *"the mass of mankind"*: Wells, *First and Last Things*, pp. 207–8.
      "the gory nurse": James, p. 1283.
      "some of the old elements": James, p. 1289.
39   *"military spirit . . . some patience"*: Nye, p. 106.
      military service *"not only educates"*: Bernhardi, p. 116.
      "up-to-date civilization": Travers, p. 267.
40   *"that we have ceased to"*: Wilkinson, p. 4.
      "unqualified desire": Bernhardi, pp. 36–37.
      "losing the psychological": Angell, p. 217.
      On the parallel evolution of militarism and pacifism, see Ceadel.

3. EUROPEANS IN A VIOLENT WORLD

42   On Zabern, see Wehler's influential essays and Schonbaum's revisionist
      views.
44   *"Hegel observes that"*: English and Townshend, p. 168.
      Herbert Spencer noted: "As fast as war ceased to be the business of life, the
      social structure produced by war . . . slowly became qualified by the social
      structure produced by industrial life." *From Freedom to Bondage* (1891), in
      Man, p. 497.

44   *"moral aversions"*: Mosca, p. 242.
      On the battle of Omdurman, see Vandervort, pp. 171–77.
45   *"But at the critical moment"*: Churchill, p. 279.
46   *"conquered on the railway"*: Churchill, p. 163.
      "campaigns rather against nature": Callwell, p. 57.
      "the pacification, restoration": Churchill, x.
      "rule by sheer violence": Arendt, *Violence*, p. 53.
      "violence administered": Arendt, *Origins*, p. 137.
47   *"In planning a war"*: Callwell, p. 40.
      British army was at war: "There was not a single year in Queen Victoria's
      long reign in which somewhere in the world her soldiers were not fight-
      ing for her and for her empire." Farwell, p. 1.
      On the French in Algeria, see Horne; on Germany in Africa, see Bley
      and Hull.
48   *"the vilest scramble"*: Hochschild, p. 4. Hochschild provides an eloquent
      indictment of Leopold's Congo and the controversy surrounding it.
49   On the use of native troops, see Vandervort, chapter 2.
50   *human costs of empire*: Farwell writes that in Britain, the price of em-
      pire "was paid, usually without qualms or regrets or very much thought."
      Wars, p. 1. See also Porter, *Absent-Minded Imperialists*.
      On the possibilities of civil war in Ireland in 1913, see Dangerfield.
51   For a dramatic example of political violence in Spain, see Ullman.
52   *"Mafiosi ensure"*: Blok, p. 8.
      Figures on Russia are from Figes, p. 202.
      rebellion swept across Romania: See Hitchens, p. 176f.
53   *Djilas spent his boyhood:* Djilas, *Land Without Justice*. On the Balkans, see
      Glenny and Mazower.
      On the Turkish-Italian war, see Romano.
54   *"a wonderful moral effect"*: Paris, p. 99.
      On the arms race, see the works by Stevenson and Hermann.
55   *"Meat is rotting"*: Trotsky, p. 272. For a military history of the Balkan
      Wars, see Hall.
      The commission report is reprinted in Carnegie Endowment, *The Other
      Balkan Wars*.
      "Every clause of international law": Carnegie Endowment, p. 13.
      "is waged not only by": Carnegie Endowment, p. 148.
56   *"has discovered the obvious"*: Carnegie Endowment, p. 17.
      "In Europe the epoch": Howard, *War*, p. 71.
58   *"The Allied and Associated"*: The Treaty of Versailles is available on the
      Avalon Project Web site.
      The first and most influential statement of Fischer's views was in the
      opening chapters of his *Griff nach der Weltmacht*, first published in 1961.
      "The nations slithered": Hamilton and Herweg, p. 38.
59   For a spirited statement on the war's domestic origins, see Mayer.
60   *Sazonov told Nicholas II:* McDonald, p. 218.

60 *"Your Excellency knows":* Mombauer, p. 151.
  "Everyone knows": Mombauer, p. 107.
  "Peace": Brooke, p. 107.
61 *"Overpowered by stormy enthusiasm":* Kershaw, vol. 1, p. 89.
62 *"a release of tension":* Jones, vol. 2, p. 171.
  "No matter what": Hamilton and Herweg, pp. 502–3.
  "reviving the sense of ": Hamilton and Herweg, pp. 502–3.
63 *"The Party is defenceless":* Joll, p. 183.
64 *"consternation was imprinted":* Verhey, p. 64.
  "Our people had": Verhey, p. 7.
  The most comprehensive study of the public's response to the outbreak
  of war is Becker.
  "from his privacy": Callois, p. 164.
65 *"Never such innocence":* Larkin, "MCMXIV," *Collected Poems*, p. 127.

4 · WAR AND REVOLUTION

69 There are a great many admirable surveys of the war. See, for example,
  Stevenson, Strachan, and Keegan.
  On St. Cyr, see LaGorce, pp. 93–94.
  On the Grenfells, see Hynes, pp. 34–35.
  "Every war is ironic": Fussell, p. 7.
70 *"Each side is inspired by":* Stromberg, p. 150.
  "What will the reward": Joll, *Intellectuals*, p. 96.
71 *"no one should be allowed":* Colin, p. 335. On the importance of the
  offensive spirit, see Howard, "Men."
  "War is an affair": Luvaas, p. 301.
  "advance with all forces": Williamson in Kennedy, p. 147.
72 *"The French army must be":* Hamilton and Herweg, p. 153. The most recent interpretation of the Schlieffen Plan is Zuber.
73 *"this is no ordinary":* Ousby, pp. 30–31.
74 *"I don't know what's to be done":* Ousby, p. 35.
  fought along the Somme: See Keegan, *Face*.
75 *"J'ai fait Verdun":* Ousby, p. 9.
  On the casualties at Verdun, see Ousby, pp. 7–8.
76 *"the acid test of a man":* Moran, p. 67.
  "will power, whereof no man has an unlimited stock": Moran, p. x.
  during the battle of Passchendaele: See Shephard, p. 53.
77 On military executions, see Oram.
  "boiling with mad rage": Gray, p. 52.
  "few soldiers, except the most noble": Bloch, *Memoirs*, p. 166.
78 *"impossible at a time":* Hardach, p. 55.
79 *"on the expectation of ":* Hardach, p. 24.
  "the economic and financial impossibility": Havighurst, p. 129.
  "Don't bother me": Doughty et al., p. 554.

79 *"a function of the engines":* Cohen, *Supreme Command,* p. 67.

80 Data on production and consumption are in Van Creveld, *Technology,* p. 175; Smith and Audoin-Rouzeau, p. 63; Van Creveld, *Command,* pp. 184–85; Challener, p. 92; Winter, *Death's Men,* p. 236; Morrow, pp. 190–91. On the mobilization of the German economy for war, see Feldman's pioneering *Army.*

81 *"We have discovered":* Porter, *War,* p. xix.
   "the subtleties of comparative advantage": D. H. Robertson quoted in Condliffe, p. 757.

83 *In Russia, where surveillance:* Holquist, p. 208.
   six hundred newsreels: Smith and Audoin-Rouzeau, p. 53.

84 *Sebastian Haffner described:* Haffner, pp. 19–21. On the Isère, see Flood.

85 Uses of airpower: Herwig, p. 191; Titmuss, p. 4; Morrow, pp. 190–91. On the effects of the British blockade, see Downes.
   "Long distance bombing": Carr, *Crisis,* p. 136.
   wartime atrocities: Horne and Kramer.

86 *"probably kindly men":* Andoin-Rouzeau and Becker, pp. 47–48.
   On the Armenian genocide, see Akçam.

87 *"all Jews and suspect individuals":* Lohr, p. 138. On the forced movement of populations, see Gattrell.
   "In the same way as I send shells": Carr, *Crisis,* p. 137.

88 *"Is this stupidity":* Paul Miliukov quoted in Treadgold, p. 119.
   "An impassable gulf": Wildman, vol. 1, p. 245.

89 *"hang (by all means hang, so people will see) . . . We are at war":* Yakovlev, pp. 20–21.

5. THE TWENTY-YEAR TRUCE

92 The most recent account of the Paris peace conference is MacMillan.

94 *"We thought we were the Romans":* Schivelbusch, p. 197.

95 *"by the aggression":* the Versailles Treaty is available on the Avalon Project Web site.
   "a moral effect": Sayer, p. 145.

96 *"I do not understand":* Balfour, p. 127. Balfour provides a useful account of imperial violence in the 1920s.
   "Sovereignty is acquired": Macfie, p. 198.

97 Figures on casualties in the Russian civil war are from Ferguson, *Pity,* pp. 391–92. For some harrowing descriptions of atrocities, see Yakovlev, p. 156.
   "The insurgent Cossack villages": Yakovlev, p. 89.
   "put an end once and for all": Figes, p. 641.

98 The classic account of the Fascists' rise to power is Lyttelton. The most recent biography of Mussolini is Bosworth.

99 *"the children of World War I":* Furet, p. 19.

"For us the war has never come to an end": Bosworth, p. 163.

The army "is that school": Odom, p. 203.

"War brings to its highest tension": Ceadel, p. 26.

100   On casualties in the war, see Ferguson, *Pity*, p. 295; Winter, *Sites*, p. 254; Smith and Audoin-Rouzeau, p. 69.

"I must be silent here": Jarausch, p. 264.

101   "to live to the end": Brittain, pp. 469–70.

On honoring the war dead, see Winter, *Sites*.

102   The symbolism of the unknown soldier is analyzed in Sherman.

104   "Somebody must suffer": Schuker, p. 5.

"an inefficient, unemployed": On the impact of Keynes, see Sontag, pp. 24–32.

"The war had won": Fussell, p. 13.

105   "The whole future": Strachan in Chickering and Förster, p. 40. Strachan's essay is a good introduction to changing views of strategy after the First World War.

106   "To assume the closing": MacIver, p. 249.

The standard work on the Kellogg-Briand Pact is Ferrell. See also Steiner for the most recent account of international relations in the twenties.

107   "reasonably safe prediction": Carr, p. 36.

"There has scarcely ever": Carr, p. 36.

"Who in Europe": *Survey of International Affairs*, 1928, p. 9.

"is surely clear": Liddell Hart, *Remaking*, p. 90.

On German losses and the problem of memorializing them, see Mosse, *Fallen*.

109   The most recent treatment of the Nazi rise to power is Evans. There is a good account of the Potempa murder on p. 296.

110   "His whole subsequent career": Canetti, *Crowds*, p. 181.

111   "Without the past war": Schivelbusch, pp. 240–41.

"There is nothing": Schivelbusch, p. 371.

112   "a low dishonest decade": Auden, "September 1, 1939," in *English Auden*, pp. 245–47. For a recent narrative of the thirties, see Large.

113   "Anything rather than war!": Weber, *Hollow Years*, p. 19.

114   "The Nazis would find": Sontag, p. 310.

Hossbach's memorandum is available on the Avalon Project Web site.

"perpetual nightmare": Sontag, p. 316.

115   "we cannot foresee": Woodward and Butler, 2nd series, vol. 19, p. 513.

"Thank goodness Austria's out of the way": Murray, p. 166.

"It is my unalterable decision": Murray, p. 173.

116   "If you have sacrificed": Wheeler-Bennett, p. 171.

117   "peace in our time": Wheeler-Bennett, p. 181.

On the immediate origins of the war, the definitive account is Watt.

"This is a sad day for all of us": Sontag, p. 381.

118   "It is a sad day for Europe": D'Este, p. 250.

118   *"the whole mobilization"*: Jackson, *Fall*, p. 145.
      "something between resolution and resignation": Weber, *Hollow Years*, p. 262.
      "The enlightenment driven away": Auden, "September 1, 1939," in *English Auden*, pp. 245–47.

6. THE LAST EUROPEAN WAR

119   *The Last European War* is the title of John Lukacs's book on the events of 1939 through 1941. There are a great many excellent studies of the Second World War. Wright's *Ordeal of Total War* is outdated in some ways but still the best survey of the war's military, political, and cultural dimensions. Weinberg's *World at Arms* provides a full and reliable narrative of political and military events.
      "not the arrival . . . greatest brutality": Rossino, p. 9. The evidence for the authenticity of these quotations is discussed in Baumgart.

120   *"The fate of Poland"*: *Survey of International Affairs, 1939–1946*, p. 245. Churchill's remarks came in the midst of a controversy over Polish participation in a parade celebrating the first anniversary of the Allied victory in Europe.
      "The very situations . . . the battle in and for itself": Aron, pp. 17–18.

122   *"incapable of thinking"*: Bloch, *Strange Defeat*, p. 36. On the fall of France, see the two books by Jackson.
      "France expiates the crime": Weber, *Hollow Years*, p. 5.
      "the spirit of enjoyment": Jackson, *France*, p. 129.

124   *"Russia will be our India"*: Overy, *Russia's War*, p. 61.
      "the greatest battle": Overy, *Russia's War*, p. 94.
      For an analysis of Germany's strategic problems, see Habeck.

125   *"So we had won after all"*: Kennedy, *Freedom*, p. 523.
      "The only thing that ever": Keegan, *Price*, pp. 225–26. The failure of the Allies' initial antisubmarine measures is described in Cohen and Gooch.

126   *"The whole question"*: Adelman, p. 135.
      Figures on weapons production: Overy, *Why the Allies Won*, pp. 331–32.

127   *the most interesting question: Why the Allies Won* is the title of Richard Overy's fine book.
      On the Sicily campaign, see D'Este, p. 438.

128   *"survives the terrible impact"*: Luttwak, *Strategy*, p. 84. The classic analysis of small-group cohesion is Shils and Janowitz. Their interpretation is challenged by Bartov. For more on the strengths of the German army, see Van Creveld, *Fighting Power*.

129   *both sides intentionally targeted*: There is a good introduction to theories of airpower in MacIsaac.
      On the destruction of Berlin, see Beevor.
      "There are no civilians": Colonel Harry F. Cunningham quoted in Schaffer, p. 142.

130   *"such a complex and interdependent fabric"*: Liddell Hart, *Remaking*, p. 106.

"the man in the street … Whatever people": Survey of International Affairs, 1932, p. 189.

"Mankind is Frankenstein": Bialer, p. 7, who incorrectly identifies the author and date of publication.

when Nazi Germany began to build: On the air war, see Biddle, *Rhetoric,* and Overy, *Air War.*

132   *at least half were citizens:* Soviet losses are recorded in Overy, *Russia's War,* p. 288.

133   *Germany's savage occupation:* The behavior of the Wehrmacht in the east is analyzed in Bartov. Dallin is the classic account of German occupation policy.

"The Third Reich's": Lieven, *Empire,* p. 68.

The most careful study of German losses is Overmans.

134   *"The gravest shortcoming":* Hastings, p. 139.

"We cannot sit off": Hastings, p. 78.

campaign of racial domination: Burleigh and Wippermann emphasize the racial dimensions of the Nazi revolution.

On Hitler's euthanasia order of October 1939, see Friedlander, p. 67.

135   *"if the international Jewish … the annihilation of":* Noakes and Pridham, vol. 3, p. 1049.

137   *"We have no interest":* Noakes and Pridham, vol. 3, p. 932.

138   On wartime Yugoslavia, see Glenny.

139   On French retribution, see Burrin, p. 459. On Europe as a whole, see the essays in Deák et al.

140   *looting, rape, and murder:* On the Soviet occupation, see Naimark, *Russians in Germany.*

142   *"expulsion [of Germans] is the method":* Claude, *Minorities,* p. 98. For an introduction to the problem of ethnic cleansing in the twentieth century, see Naimark, *Fires.*

On displaced persons, see Marrus.

viewed as traitors: For the Soviet persecution of "collaborators," see Yakovlev.

143   *"You must resign yourselves":* Adelman, p. 253. On the importance of wartime service in the Soviet Union, see Weiner.

"which would be useful": Connelly, p. 230. For the question of war memorials, see Winter, *Sites,* and Mosse, *Fallen,* both of which compare the two world wars.

144   *One of the best collections:* In Bessel and Schumann, see especially the essay by Pieter Lagrou.

7. THE FOUNDATIONS OF THE POSTWAR WORLD

148   The best single book on the Cold War is Soutou. In English, Powaski provides a good brief summary.

149   On American military planning, see Kennedy, *Freedom,* pp. 486–87.

150   For Truman on the Nazis and Soviets, see Powaski, p. 48.
      "Everyone imposes": Naimark, "Stalin," p. 36. Naimark provides a clear
      and balanced analysis of Soviet postwar strategy.

152   *"I believe that it must"*: Truman's speech is available on the Avalon Project
      Web site.

154   *evolution of the German problem*: Trachtenberg, "Making," p. 103. On the
      centrality of the German question, see the books by Trachtenberg and
      Eisenberg.

156   *"There is no longer any difference"*: Documents on International Affairs, 1949–
      1950, p. 66.

157   *"limited liability"*: Reynolds, "1940," p. 349. The following two paragraphs
      are based on this superb article. See also Hitchcock, *France*.

158   *Schuman's plan "was bold, simple"*: Gillingham, *Coal*, p. 231.
      Schuman's speech is available on the Avalon Project Web site.
      "the word 'Europe'": Gillingham, *Coal*, p. 231.

159   On the impact of the Korean War, see Stueck.
      The NATO Treaty is available on the Avalon Project Web site.

160   *"there will be German soldiers but no German army"*: Schuman, "France,"
      p. 355.
      "obviously imposes on the United States": Ruane, p. 102.
      "keep the Russians out": As is the case with many famous quotations, there
      is considerable controversy over what Ismay meant. On the evolution of
      NATO, see Cleveland.
      On the ambiguities of German sovereignty, see Herbst.

161   On the evolution of European integration, see Milward, *European Rescue*, and Gillingham, *European Integration*.
      The classic "rational choice" analysis of the European project is Moravcsik.

162   *"produce collapse"*: Reynolds, "Security 'Lessons,'" p. 311.

163   *"Unless we bring the nuclear deterrent into play"*: Survey of International
      Affairs, 1959–1960, p. 103.
      The best introduction to the history of nuclear strategy is Freedman.

164   *"the symbol and even the characteristic aspect"*: Survey of International Affairs,
      1961, p. 45.

165   *"Nowhere in the world"*: Craig, "Militarization," p. 105.

167   *"the greatest transfer of power"*: Herbst in Rotberg, p. 312. There is an excellent account of this process in Abernethy.
      On the Algerian War, see Horne. There is an analysis of French opinion
      in Lustick.

169   *"the winds of change"*: Connelly, *Revolution*, pp. 279–80.
      "the Mecca of the revolutionaries": Connelly, *Revolution*, pp. 279–80.

170   *"Imperialism has ceased to bring"*: Connelly, *Revolution*, p. 31.
      "might count for nothing": Lustick, p. 88.
      "Algeria costs us": Lustick, p. 324. Wall has a critical account of de Gaulle's
      policy.
      "They had been the centres": Fieldhouse, p. 394.

8. THE RISE OF THE CIVILIAN STATE

172   *"in which the specialists":* Lasswell revised his 1937 article four years later, and it was this version that remained extremely influential for the next two decades. See Lasswell, *Essays.*

173   *explosive economic growth:* Milward, *Reconstruction.* Data from Judt, p. 338f.

174   *"Foreign trade is":* Garton Ash, *Europe's Name,* p. 244.

175   *Postwar public opinion surveys:* Inglehart, p. 49.
      "The general effect of the war": Rosanvallon, p. 28.
      "spirit of brotherhood": Rosanvallon, p. 28.
      the British government's expenses: Ferguson, *Cash,* pp. 100–101. On other European states, see the data in Flora.

176   *Military expenditures:* Stockholm International Peace Research Institute, *Yearbook 1976,* pp. 152–56, and *Yearbook 1985,* p. 281. On Britain: Harries-Jenkins, pp. 11, 108. On the Netherlands: Olivier and Teitler, pp. 86–87. The best analysis of a national military budget is in Martin, *Warriors.*
      "The budget reveals": Genschel and Uhl, p. 92.
      "strong defense forces": Inglehart, p. 49. British opinion: Shaw, p. 123f.

177   *"is always a significant":* Kiernan, p. 141.
      Conscription in Britain: Scott, *Conscription.* In France: Martin, *Warriors.*

178   *"civilization":* Shaw, p. 74, and also Olivier and Teitler. For developments in other European states, see the works by Kelleher, Burk, and Flynn ("Conscription" and *Conscription*) and the essays in Harries-Jenkins and in Moskos, Williams, and Segal.
      On alternative service, see Bartjes.

179   *"a great community":* Hedetoft, p. 281.
      "death was no longer": Howard, *Invention,* p. 100.
      how little space these issues were given: There are, for example, no references to "army," "conscription," "military," "security," or "war" in Gabriel Almond and Sidney Verba's classic study, *Civic Culture,* published in 1963. Similarly, military institutions are largely ignored in the influential collection of essays *A New Europe?,* which first appeared in *Daedalus* in winter 1964.

181   *"Dissidence, involving":* Suri, p. 2.

182   On violence in 1968: Tilly, *Collective Violence,* pp. 63 (on France), 71 (on Germany).

183   Data on political terrorism can be found in Chalk.

184   Data on deaths from political violence in Ireland can be found in O'Day, p. 18.
      "an acceptable level": Beckett, p. 15.

185   On Greece, see Clogg, chapters 5 and 6.
      There is a good account of the Portuguese revolution in Maxwell.

186   On Spain, see the works by Preston and by Carr and Fusi.
      "When the revolution occurred": Maxwell, p. 66.

187   *"Nations prefer to live prosaically"*: Maxwell, p. 30.
188   Data on arrests in 1945 are from Mazower, *Dark Continent*, p. 265.
      "repression without annihilation": Todorov, *Hope*, p. 43.
      "post-totalitarian": Schell, p. 195.
189   There is an excellent account of the Khrushchev era in Taubman.
      "a specific form of the international class struggle": Odom, p. 12.
      "a permanent wartime economy": Odom, p. 63.
191   *Andropov and Czech repression:* Soutou, p. 462.
      "minor mishap": Soutou, p. 482. Garton Ash, *Europe's Name*, provides a
      powerful analysis of the assumptions behind European approaches to the
      East.
      Data on indebtedness are from Maier, p. 63.
192   *"To believe in overthrowing"*: Schell, p. 191. For new styles of peaceful pro-
      test, see Kenney.
      "to live within the truth": Schell, p. 195.
193   *"historically abnormal"*: *Economist*, February 2, 1985, pp. 14–15.
194   *"Only later was it easy"*: Gorbachev, p. 39.
      "force and the threat of force": Gorbachev's speech is available on the Cold
      War Files Web site.
195   *end of empire had very different results in East and West:* Mazower, *Dark
      Continent*, pp. 376–77.
      dissolution of empire: Dallin, "Causes," gives a balanced assessment of the
      Soviet collapse. Kotkin correctly emphasizes its peaceful character.

9. WHY EUROPE WILL NOT BECOME A SUPERPOWER

200   *"the distribution and character of military power"*: Mearsheimer, p. 6.
      "stark simplicities": Waltz, p. 44.
      "relearn their old great power roles": Waltz, p. 72.
      "The prospects for major": Mearsheimer, p. 6. See also Layne.
201   *"Today sovereignty means"*: Speech of October 3, 1990, marking German
      unification. Weizsäcker, p. 194.
      "the profound political": The "New Strategic Concept" is available on the
      NATO Web site.
202   For the context of Kiep's book and the underlying problems of the Amer-
      ican alliance, see Haftendorn.
      Texts of these treaties are available on the European Union Web site.
204   *"The hour of Europe"*: Silber and Little, p. 159. As with other famous quo-
      tations, the source is difficult to trace. Garton Ash is skeptical about the
      wording but not the sentiment. *Free World*, p. 255, note 31.
205   *"The conflict in Bosnia was a product"*: Mazower, *Balkans*, p. 147. For a
      sharply critical analysis of British policy, see Simms.
206   *"do not encroach"*: Millon, p. 13.
      "a European Security and Defense Identity": Press communiqué, NATO min-
      isterial meeting, Berlin, June 3, 1996. See Cogan, p. 84.

Text of the Treaty of Amsterdam is available on the European Union Web site.

207   On the Cologne summit, see Cogan, pp. 110–13.
      On the first Gulf War, see Freedman and Karsh.
      On the war in Kosovo, see the essays in Bacevich and Cohen, and for its impact on Europe, see Cogan, p. 104f.

209   *"is not comparable":* Ikenberry, p. 270.

210   For more on the "Revolution in Military Affairs," see the works by Sloan, Hirst, and Latham. For its impact on American strategic thinking, see Bacevich, *New American Militarism,* pp. 166–73.
      On the initial European response to the events of September 11, see Gordon and Shapiro. On NATO's response, see Deighton and also Gordon, "NATO."

211   *"war on terrorism . . . arouses an expectation":* Gordon and Shapiro, p. 61. European opinion tended to support this view: Gordon, "NATO," p. 94. See also Freedman, "Third World War," and Katzenstein, "Same War —Different Views."
      "would be made up": Gordon, "NATO," p. 92. When the United States began hostilities in Afghanistan, it did not bother to inform the president of the European Union.

212   *"deconstructing the old":* New York Times, November 4, 2002.
      One word was conspicuously absent: The Prague communiqué is available on the NATO Web site.

213   *Many European governments:* See Gordon and Shapiro, and Garton Ash, *Free World.* On the roots and implications of anti-Americanism, see Markovits.
      disregard for consultation: "The diplomacy of the Iraq crisis," Garton Ash observed, "was a case study, on all sides, of how not to run the world." *Free World,* p. 37. For a summary of events, see Gordon and Shapiro.

214   *"crisis of credibility":* Nicholas Burns quoted in Gordon and Shapiro, p. 138.

215   *more and more European units:* In November 2006, NATO and its non-alliance partners had 32,000 troops in Afghanistan. Many of them operated under restrictive orders that limited their combat role. *New York Times,* November 27, 2006.
      "Europe does not need": Haseler, pp. 15–16.

216   *"to develop a strategic culture":* The statement is available on the NATO Web site.
      "Bosnia is a demonstration": Economist, March 19, 2005, p. 60.

217   For estimates on the cost of the Rapid Response Force, see Wolf and Zycher.
      while the cost of military hardware: On the contrast between rising military costs and shrinking budgets, see Cusack and the data in International Institute for Strategic Studies, *The Military Balance* (2004).

218   *largest economic bloc:* On European economic power, see Zielonka, p. 91.

220   *"the European idea is empty":* Aron, p. 316.

EPILOGUE: THE FUTURE OF THE CIVILIAN STATE

222   *"less and less logical":* Cogan, 134. For contrasting views on the future of
      European security, see Sangiovanni and Morgan.
223   *"non-war community":* Wæver, p. 69.
      "post-heroic age": Luttwak, "Post-Heroic."
224   *"a living laboratory":* Haas, p. 4.
      "Europe has become the new 'city upon a hill'": Rifkin, p. 358.
225   *"soft power":* On the international importance of Europe's soft power, see
      the works by Leonard and Zielonka.
      "desecuritization": Wæver, p. 69.
      "One of the most vexing questions": Keohane and Nye, p. 6. On the geo-
      graphical distribution of violence, see O'Louglin in Flint.
226   *it will not be an easy matter:* In January 2006, for example, the European
      Court of Human Rights ruled that Turkey had breached international
      law by giving repeated prison sentences to a conscientious objector. Os-
      man Murat Ulke served seven hundred days in jail for refusing to per-
      form his military service. Agence France Presse, January 25, 2006.
      "a gated community": Garton Ash, *Free World,* p. 149.

Bibliography

Abernethy, David. *The Dynamics of Global Dominance: European Overseas Empires, 1415–1980.* New Haven: Yale University Press, 2000.

Adelman, Jonathan. *Prelude to the Cold War: Tsarist, Soviet, and U.S. Armies in the Two World Wars.* Boulder, Colo.: L. Rienner, 1988.

Ahmann, R., A. M. Birke, and M. Howard, eds. *The Quest for Stability: Problems of West European Security, 1918–1957.* Oxford: Oxford University Press, 1993.

Akçam, Tamer. *A Shameful Act: The Armenian Genocide and the Question of Turkish Responsibility.* New York: Metropolitan, 2006.

Angell, Norman. *The Great Illusion: A Study of the Relation of Military Power to National Advantage.* 4th rev. and enl. ed. London: G. P. Putnam's Sons, 1913.

Arendt, Hannah. *The Origins of Totalitarianism.* New York: Harcourt, Brace, 1951.

———. *On Violence.* New York: Harcourt, Brace, Jovanovich, 1970.

Aron, Raymond. *The Century of Total War.* Boston: Beacon Press, 1955.

Auden, W. H. *The English Auden: Poems, Essays, and Dramatic Writings, 1927–1939.* New York: Random House, 1977.

Audoin-Rouzeau, Stéphane, and Annette Becker. *Understanding the Great War.* Translated by Catherine Temerson. New York: Hill and Wang, 2002.

Avalon Project at Yale Law School. Documents in Law, History, and Diplomacy. www.yale.edu/lawweb/avalon/avalon.htm.

Bacevich, Andrew J. *The New American Militarism: How Americans Are Seduced by War.* New York: Oxford University Press, 2005.

———, and Eliot Cohen, eds. *War over Kosovo: Politics and Strategy in a Global Age.* New York: Columbia University Press, 2001.

Balfour, Sebastian. *Deadly Embrace: Morocco and the Road to the Spanish Civil War.* New York: Oxford University Press, 2002.

Bartjes, Heinz. "Der Zivildienst als die modernere 'Schule der Nation'?" In *Von der Kriegskultur zur Friedenskultur? Zum Mentalitätswandel in Deutschland seit 1945,* edited by Thomas Kühne, 128–43. Münster: Lit, 2000.

Bartov, Omer. *The Eastern Front, 1941–1945: German Troops and the Barbarisation of Warfare.* New York: Palgrave, 2001.

Baumgart, Winfried. "Zur Ansprache Hitlers vor den Führen der Wehrmacht am 22. August 1939." *Vierteljahrshefte für Zeitgeschichte* 16 (April 1968): 120–49.

Beauvoir, Simone de. *The Second Sex.* New York: Vintage, 1989.

Becker, Jean-Jacques. *1914: comment les Français sont entrés dans la guerre.* Paris: Presses de la Fondation Nationale des Sciences Politiques, 1977.

Beckett, Ian. *Insurgency in Iraq: An Historical Perspective.* Carlisle, Pa.: Strategic Studies Institute, Army War College, January 2005.

Beevor, Antony. *Berlin: The Downfall, 1945.* London: Viking, 2002.

Bernhardi, Friedrich von. *Germany and the Next War.* Translated by Allen H. Powles. New York: Longmans, Green, 1914. Originally published in 1912.

Bessel, Richard, and Dirk Schumann, eds. *Life After Death: Approaches to a Cultural and Social History of Europe During the 1940s and 1950s.* New York: Cambridge University Press, 2003.

Bialer, Uri. *Shadow of the Bomber: The Fear of Air Attack and British Politics, 1932–1939.* London: Royal Historical Society, 1980.

Biddle, Tami Davis. *Rhetoric and Reality in Air Warfare: The Evolution of British and American Ideas About Strategic Bombing, 1914–1945.* Princeton, N.J.: Princeton University Press, 2002.

Bley, Helmut. *Kolonialherrschaft und Sozialstruktur in Deutsch-Südwestafrika, 1894–1914.* Hamburg: Leibniz, 1968.

Bloch, Jean de. *The Future of War in Its Technical, Economic, and Political Relations.* Translated by R. C. Long. Boston: World Peace Foundation, 1899.

Bloch, Marc. *Memoirs of War, 1914–1915.* Ithaca, N.Y.: Cornell University Press, 1980.

———. *Strange Defeat: A Statement of Evidence.* New York: Norton, 1999. Written in 1940.

Blok, Anton. *The Mafia of a Sicilian Village, 1860–1960: A Study of Violent Peasant Entrepreneurs.* New York: Harper & Row, 1974.

Bond, Brian. *War and Society in Europe, 1870–1970.* Montreal: McGill-Queen's University Press, 1998.

Bosworth, R.J.B. *Mussolini.* London: Arnold, 2002.

Brittain, Vera. *Testament of Youth.* New York: Wideview, 1978. Written in 1933.

Brooke, Rupert. *The Collected Poems.* New York: John Lane, 1921.

Burk, James. "The Decline of Mass Armed Forces and Compulsory Military Service." *Defense Analysis* 8 (1992): 45–59.

Burleigh, Michael, and Wolfgang Wippermann. *The Racial State: Germany, 1933–1945.* Cambridge: Cambridge University Press, 1991.

Burrin, Philippe. *France Under the Germans, 1940–1944.* New York: Norton, 1996.

Callois, Roger. *Man and the Sacred.* Westport, Conn.: Greenwood Press, 1980.

Callwell, C. E. *Small Wars: Their Principles and Practice.* 3rd ed. London: General Staff, War Office, 1909. Written in 1899.

Canetti, Elias. *Crowds and Power.* New York: Seabury, 1978.

Carnegie Endowment for International Peace. *The Other Balkan Wars: A 1913*

Carnegie Endowment Inquiry in Retrospect. Washington, D.C.: Carnegie Endowment for International Peace, 1993.

Carr, E. H. *The Twenty Years' Crisis, 1919–1939: An Introduction to the Study of International Relations.* New York: St. Martin's Press, 1946.

Carr, Raymond, and Juan Pablo Fusi. *Spain: Dictatorship to Democracy.* 2nd ed. London: Allen and Unwin, 1981.

Ceadel, Martin. *Thinking About War and Peace.* Oxford: Oxford University Press, 1987.

Ceva, Lucio. *Storia della forze armate in Italia.* Torino: UTET, 1999.

Chalk, Peter. *West European Terrorism and Counter-Terrorism: The Evolving Dynamic.* London: Macmillan, 1996.

Challener, Richard. *The French Theory of the Nation in Arms, 1866–1939.* New York: Columbia University Press, 1955.

Chesney, G. T. "The Battle of Dorking: Reminiscences of a Volunteer" (1871). In *The Tale of the Next Great War, 1871–1914,* edited by I. F. Clark, 3–48. Syracuse, N.Y.: Syracuse University Press, 1996.

Chickering, Roger, and Stig Förster, eds. *The Shadows of Total War: Europe, East Asia, and the United States, 1919–1939.* New York: Cambridge University Press, 2003.

Churchill, W. S. *The River War: An Account of the Reconquest of the Sudan.* London: Eyre and Spottiswoode, 1951. Written in 1899.

Claude, Inis. *National Minorities: An International Problem.* Cambridge, Mass.: Harvard University Press, 1955.

Cleveland, Harald. *The Atlantic Idea and Its European Rivals.* New York: McGraw-Hill, 1966.

Clogg, Richard. *A Concise History of Greece.* 2nd ed. Cambridge: Cambridge University Press, 2002.

Cogan, Charles. *The Third Option: The Emancipation of European Defense, 1989–2000.* Westport, Conn.: Praeger, 2001.

Cohen, Eliot. *Supreme Command: Soldiers, Statesmen, and Leadership in Wartime.* New York: Free Press, 2002.

———, and John Gooch. *Military Misfortunes: The Anatomy of Failure in War.* New York: Free Press, 1990.

Cold War Files. www.coldwarfiles.org.

Colin, Jean-Lambert-Alphonse. *The Transformations of War.* Westport, Conn.: Greenwood Press, 1977. Originally published in 1912.

Condliffe, J. B. *The Commerce of Nations.* New York: Norton, 1955.

Connelly, Mark. *The Great War: Memory and Ritual Commemoration in the City and East London.* Rochester, N.Y.: Royal Historical Society/Boydell, 2002.

Connelly, Matthew. *A Diplomatic Revolution: Algeria's Fight for Independence and the Origins of the Post–Cold War Era.* New York: Oxford University Press, 2002.

Craig, Gordon. *The Politics of the Prussian Army, 1640–1945.* Oxford: Clarendon Press, 1955.

———. *Königgrätz 1866: Eine Schlacht macht Weltgeschichte.* Augsburg: Bechtermünz, 1997.

———. "The Militarization of Europe, 1945–1986." In *The Militarization of the Western World,* edited by John Gillis. New Brunswick, N.J.: Rutgers University Press, 1989.

Cusack, Thomas. "Sinking Budgets and Ballooning Prices: Recent Developments Connected to Military Spending." In *Discussion Papers.* Wissenschaftszentrum Berlin, April 2006.

Dallin, Alexander. *German Rule in Russia, 1941–1945.* 2nd ed. London: Macmillan, 1981.

———. "Causes of the Collapse of the USSR." *Post-Soviet Affairs* 8, no. 4 (1992): 279–302.

Dangerfield, George. *The Strange Death of Liberal England.* New York: Capricorn, 1935.

Deák, Istvan, Jan T. Gross, and Tony Judt, eds. *The Politics of Retribution in Europe: World War II and Its Aftermath.* Princeton, N.J.: Princeton University Press, 2000.

Deighton, Anne. "The Eleventh of September and Beyond: NATO." In *Superterrorism: Policy Responses,* edited by Lawrence Freedman, 119–34. Malden, Mass.: Blackwell, 2002.

Delbrück, Hans. "Zukunftskrieg und Zukunftsfriede." *Preussische Jahrbücher* 96 (1899): 203–29.

D'Este, Carlo. *Eisenhower: A Soldier's Life.* New York: Henry Holt, 2002.

Djilas, Milovan. *Land Without Justice.* New York: Harcourt, Brace, Jovanovich, 1958.

Documents on International Affairs. 19 vols. London: Oxford University Press, 1929–1973.

Dorpalen, Andreas. *Heinrich von Treitschke.* New Haven: Yale University Press, 1957.

Doughty, Robert, et al. *Warfare in the Western World: Military Operations.* 2 vols. Lexington, Mass.: D. C. Heath, 1996.

Downes, Alexander. "Desperate Times, Desperate Measures: The Causes of Civilian Victimization in War." *International Security* 30, no. 4 (Spring 2006): 152–95.

Dülffer, Jost. *Regeln gegen den Krieg? die Haager Friedenskonferenz von 1899 und 1907 in der internationalen Politik.* Berlin: Ullstein, 1981.

Dunlop, John K. *The Development of the British Army, 1899–1914.* London: Methuen, 1938.

Eisenberg, Carolyn. *Drawing the Line: The American Decision to Divide Germany, 1944–1949.* Cambridge: Cambridge University Press, 1996.

English, Richard, and Charles Townshend, eds. *The State.* London: Routledge, 1999.

European Union Web site: http://europa.eu.

Evans, Richard. *The Coming of the Third Reich.* New York: Penguin, 2004.

Farwell, Byron. *Queen Victoria's Little Wars.* New York: Harper & Row, 1972.

Feldman, Gerald. *Army, Industry, and Labor in Germany, 1914–1918.* Princeton, N.J.: Princeton University Press, 1966.

Ferguson, Niall. *The Pity of War.* London: Penguin, 1998.

———. *The Cash Nexus: Money and Power in the Modern World, 1700–2000.* New York: Basic Books, 2001.

Ferrell, Robert. *Peace in Their Time: The Origins of the Kellogg-Briand Pact.* New Haven: Yale University Press, 1952.

Fieldhouse, D. K. *The West and the Third World: Trade, Colonialism, Dependence, and Development.* Oxford: Blackwell, 1999.

Figes, Orlando. *A People's Tragedy: The Russian Revolution, 1891–1924.* London: Jonathan Cape, 1996.

Flint, Colin, ed. *The Geography of War and Peace: From Death Camps to Diplomats.* New York: Oxford University Press, 2005.

Flood, P. J. *France 1914–1918: Public Opinion and the War Effort.* New York: Macmillan, 1990.

Flora, Peter, et al. *State, Economy, and Society in Western Europe, 1815–1975.* 2 vols. Frankfurt: Campus, 1983.

Flynn, George. *Conscription and Democracy: The Draft in France, Great Britain, and the United States.* Westport, Conn.: Greenwood Press, 2002.

———. "Conscription and Equity in Western Democracies, 1940–1975." *Journal of Contemporary History* 33, no. 1 (1998): 5–20.

Freedman, Lawrence. *The Evolution of Nuclear Strategy.* New York: Macmillan, 1983.

———. "The Third World War?" *Survival* 43, no. 4 (2001–2): 61–88.

———, and Efraim Karsh. *The Gulf Conflict, 1990–1991: Diplomacy and War in the New World Order.* Princeton, N.J.: Princeton University Press, 1993.

Frevert, Ute. *Die kasernierte Nation: Militärdienst und Zivilgesellschaft in Deutschland.* Munich: Beck, 2001.

Friedlander, Henry. *The Origins of Nazi Genocide: From Euthanasia to the Final Solution.* Chapel Hill: University of North Carolina Press, 1995.

Furet, François. *The Passing of an Illusion: The Idea of Communism in the Twentieth Century.* Chicago: University of Chicago Press, 1999.

Fussell, Paul. *The Great War and Modern Memory.* London: Oxford University Press, 1975.

Garton Ash, Timothy. *In Europe's Name: Germany and the Divided Continent.* New York: Random House, 1993.

———. *Free World: America, Europe, and the Surprising Future of the West.* New York: Random House, 2004.

Gat, Azar. *War in Human Civilization.* Oxford: Oxford University Press, 2006.

Gatrell, Peter. *A Whole Empire Walking: Refugees in Russia During World War I.* Bloomington: Indiana University Press, 1999.

Genschel, Philipp, and Suzanne Uhl. "Der Steuerstaat und die Globalisierung." In *Transformationen des Staates?*, edited by S. Leibfried and M. Zürn, 92–122. Frankfurt: Suhrkamp, 2006.

Gillingham, John. *Coal, Steel, and the Rebirth of Europe, 1945–1955: The Germans and the French from Ruhr Conflict to European Community*. New York: Cambridge University Press, 1991.

———. *European Integration, 1950–2003: Superstate or New Market Economy?* Cambridge: Cambridge University Press, 2003.

Glenny, Misha. *The Balkans, 1804–1999: Nationalism and the Great Powers*. London: Granta, 1999.

Gooch, John. *Armies in Europe*. London: Routledge and Kegan Paul, 1980.

———. *Army, State, and Society in Italy, 1870–1915*. Basingstoke: Macmillan, 1989.

Gorbachev, Mikhail. *Perestroika: New Thinking for Our Country and the World*. New York: Harper & Row, 1988.

Gordon, P. H. "NATO after 11 September." *Survival* 43, no. 4 (2001–2): 89–106.

———, and Jeremy Shapiro. *Allies at War: America, Europe, and the Crisis over Iraq*. New York: McGraw-Hill, 2004.

Gosewinkel, Dieter. *Einbürgern und Ausschliessen: die Nationalisierung der Staatsangehörigkeit vom Deutschen Bund bis zur Bundesrepublik Deutschland*. Göttingen: Vandenhoeck and Ruprecht, 2001.

Gray, J. Glenn. *Warriors: Reflections on Men in Battle*. New York: Harcourt, Brace, 1959.

Haas, Ernst. *The Uniting of Europe: Political, Social, and Economic Forces, 1950–1957*. Stanford, Calif.: Stanford University Press, 1958.

Habeck, Mary. *Storm of Steel: The Development of Armor Doctrine in Germany and the Soviet Union, 1919–1939*. Ithaca, N.Y.: Cornell University Press, 2003.

Habermas, Jürgen. *Der gespaltene Westen*. Frankfurt: Suhrkamp, 2004.

Haffner, Sebastian. *Geschichte eines Deutschen: Erinnerungen, 1914–1933*. Stuttgart: Deutsche Verlags-Anstalt, 2000.

Haftendorn, Helga. *Eine schwierige Partnerschaft: Bundesrepublik Deutschland und USA im Atlantischen Bündnis*. Berlin: Quorum, 1988.

Hall, Richard. *The Balkan Wars, 1912–1913: Prelude to the First World War*. London: Routledge, 2000.

Hamilton, Richard, and Holger Herwig, eds. *The Origins of World War I*. New York: Cambridge University Press, 2002.

Hardach, Gerd. *The First World War, 1914–1918*. Berkeley: University of California Press, 1977.

Harries-Jenkins, Gwyn, ed. *Armed Forces and the Welfare Societies: Challenges in the 1980s*. New York: St. Martin's Press, 1983.

Haseler, Stephen. "Rethinking NATO: A European Declaration of Independence." *The Federal Trust, European Essay No. 26*, 2003.

Hastings, Max. *Armageddon: The Battle for Germany*. New York: Knopf, 2004.

Havighurst, Alfred F. *Twentieth-Century Britain*. 2nd ed. New York: Harper & Row, 1966.

Hedetoft, Ulf. "National Identity and Mentalities of War in Three EC Countries." *Journal of Peace Research* 30, no. 3 (1993): 281–300.

Herbst, Ludolf. "Wie souverän ist die Bundesrepublik?" In *Sieben Fragen an die Bundesrepublik, Vorträge aus dem Institut für Zeitgeschichte,* edited by W. Benz, 72–90. Munich: Deutsches Taschenbuch, 1989.

Herrmann, David. *The Arming of Europe and the Making of the First World War.* Princeton, N.J.: Princeton University Press, 1996.

Herwig, Holger. *The First World War: Germany and Austria-Hungary, 1914–1918.* London: Arnold, 1997.

Hirst, Paul. *War and Power in the 21st Century: The State, Military Conflict, and the International System.* Oxford: Polity, 2001.

Hitchcock, William. *France Restored: Cold War Diplomacy and the Quest for Leadership in Europe, 1944–1955.* Chapel Hill: University of North Carolina Press, 1998.

Hitchens, Keith. *Rumania, 1866–1947.* Oxford: Clarendon Press, 1994.

Hobson, J. A. *Imperialism: A Study.* Ann Arbor: University of Michigan Press, 1965. Originally published in 1902.

Hochschild, Adam. *King Leopold's Ghost: A Story of Greed, Terror, and Heroism in Colonial Africa.* Boston: Houghton Mifflin, 1999.

Holquist, Peter. *Making War, Forging Revolution: Russia's Continuum of Crises, 1914–1921.* Cambridge, Mass.: Harvard University Press, 2002.

Horne, Alistair. *A Savage War of Peace: Algeria, 1954–1962.* New York: Penguin, 1987.

Horne, John, and Alan Kramer. *German Atrocities 1914: A History of Denial.* New Haven: Yale University Press, 2001.

Howard, Michael. *War and the Liberal Conscience.* London: Temple Smith, 1978.

———. *The Invention of Peace.* New Haven: Yale University Press, 2000.

———. "Men Against Fire: The Doctrine of the Offensive in 1914." In *Makers of Modern Strategy: From Machiavelli to the Nuclear Age,* edited by Peter Paret, 510–26. Princeton, N.J.: Princeton University Press, 1986.

Hull, Isabel V. *Absolute Destruction: Military Culture and the Practices of War in Imperial Germany.* Ithaca, N.Y.: Cornell University Press, 2005.

Hynes, Samuel. *The Soldier's Tale: Bearing Witness to Modern War.* New York: Allen Lane/Penguin, 1997.

Ikenberry, John, ed. *America Unrivaled: The Future of the Balance of Power.* Ithaca, N.Y.: Cornell University Press, 2002.

Inglehart, Ronald. *The Silent Revolution: Changing Values and Political Styles Among Western Publics.* Princeton, N.J.: Princeton University Press, 1977.

International Institute for Strategic Studies. *The Military Balance.* London: International Institute for Strategic Studies, 1954–2004.

Jackson, Julian. *The Fall of France: The Nazi Invasion of 1940.* Oxford: Oxford University Press, 2003.

———. *France: The Dark Years, 1940–1944.* Oxford: Oxford University Press, 2003.

James, William. "The Moral Equivalent of War." In *Writings, 1902–1910*, 1281–93. New York: Viking, 1987.

Janowitz, Morris. "Military Institutions and Citizenship in Western Societies." In *The Military and the Problem of Legitimacy*, edited by Gwyn Harries-Jenkins and Jacques Van Doorn, 77–92. Beverly Hills, Calif.: Sage Publications, 1976.

Jarausch, Konrad. *The Enigmatic Chancellor: Bethmann Hollweg and the Hubris of Imperial Germany.* New Haven: Yale University Press, 1973.

Joas, Hans. *War and Modernity.* Cambridge: Polity, 2003.

Joll, James. *Intellectuals in Politics: Three Biographical Essays.* London: Weidenfeld and Nicolson, 1960.

———. *The Origins of the First World War.* London: Longman, 1984.

Jones, Ernest. *The Life and Work of Sigmund Freud*, vol. 2, *Years of Maturity, 1901–1919.* New York: Basic Books, 1955.

Jouvenel, Bertrand de. *Power: The Natural History of Its Growth.* London: Hutchinson, 1948.

Kagan, Robert. *Of Paradise and Power: America and Europe in the New World Order.* New York: Vintage, 2004.

———. "Power and Weakness." *Policy Review* 113 (2002): 1–21.

Katzenstein, Peter. "Same War—Different Views: Germany, Japan, and Counterterrorism." *International Organization* 57, no. 4 (Fall 2003): 731–60.

Keegan, John. *The Face of Battle.* New York: Viking, 1976.

———. *The Price of Admiralty: The Evolution of Naval Warfare.* New York: Viking, 1989.

———. *War and Our World.* New York: Vintage, 1998.

———. *The First World War.* New York: Knopf, 1999.

Kelleher, Catherine. "Mass Armies in the 1970s: The Debate in Western Europe." *Armed Forces and Society* 1 (November 1978): 3–30.

Kennedy, David M. *Freedom from Fear: The American People in Depression and War, 1929–1945.* New York: Oxford University Press, 1999.

Kennedy, Paul, ed. *War Plans of the Great Powers, 1880–1914.* London: Unwin, 1985.

Kenney, Padraic. *A Carnival of Revolutions: Central Europe, 1989.* Princeton, N.J.: Princeton University Press, 2002.

Keohane, Robert, and Joseph Nye. *After the Cold War: International Institutions and Strategies in Europe, 1989–1991.* Cambridge, Mass.: Harvard University Press, 1983.

Kershaw, Ian. *Hitler.* 2 vols. New York: Norton, 1998.

Kiep, Walter. *Good-bye Amerika, was dann? Der deutsche Standpunkt im Wandel der Weltpolitik.* Stuttgart: Seewald, 1972.

Kiernan, Victor. "Conscription and Society in Europe Before the War of 1914–1918." In *War and Society*, edited by M.R.D. Foot, 141–58. New York: Barnes and Noble, 1973.

Kotkin, Stephen. *Armageddon Averted: The Soviet Collapse, 1970–2000.* New York: Oxford University Press, 2001.

Kühne, Thomas, ed. *Von der Kriegskultur zur Friedenskultur? Zum Mentalitätswandel in Deutschland seit 1945.* Münster: Lit, 2000.

La Gorce, Paul Marie de. *The French Army: A Military-Political History.* New York: G. Braziller, 1963.

Lagroll, Pieter. "The Nationalization of Victimhood: Selective Violence and National Grief in Western Europe, 1940–1960." In *Life After Death: Approaches to a Cultural and Social History of Europe During the 1940s and 1950s,* edited by Richard Bessel and Dirk Schumann. New York: Cambridge University Press, 2003.

Laity, Paul. *The British Peace Movement, 1870–1914.* New York: Oxford University Press, 2002.

Large, David Clay. *Between Two Fires: Europe's Path in the 1930s.* New York: Norton, 1990.

Larkin, Philip. *Collected Poems.* New York: Farrar, Straus and Giroux, 1989.

Lasswell, Harold. *Essays on the Garrison State.* New Brunswick, N.J.: Transaction, 1997.

Latham, Andrew. "Warfare Transformed: A Braudelian Perspective on the 'Revolution in Military Affairs.'" *European Journal of International Relations* 8 no. 2 (2002): 231–66.

Layne, Christopher. "The Unipolar Illusion: Why New Great Powers Will Rise." *International Security* 17, no. 4 (1993): 5–51.

Leibfried, Stephan, and Michael Zürn, eds. *Transformationen des Staates?* Frankfurt: Suhrkamp, 2006.

Leonard, Mark. *Why Europe Will Rule the 21st Century.* London: Fourth Estate, 2005.

Lepsius, J., et al., eds. *Die grosse Politik der europäischen Kabinette, 1871–1914.* Vol. 15. Berlin: Deutsche Verlagsgesellschaft für Politik und Geschichte, 1924.

Liddell Hart, B. H. *The Remaking of Modern Armies.* Boston: Little, Brown, 1928.

Lieven, Dominic. *Empire: The Russian Empire and Its Rivals.* New Haven: Yale University Press, 2000.

Lindberg, Tod, ed. *Beyond Paradise and Power: Europe, America, and the Future of a Troubled Partnership.* New York: Routledge, 2005.

Lohr, Eric. *Nationalizing the Russian Empire: The Campaign Against Enemy Aliens During World War I.* Cambridge, Mass.: Harvard University Press, 2003.

Lukacs, John. *The Last European War: September 1939/December 1941.* Garden City, N.Y.: Anchor Press, 1976.

Lustick, Ian. *Unsettled States, Disputed Lands: Britain and Ireland, France and Algeria, Israel and West Bank–Gaza.* Ithaca, N.Y.: Cornell University Press, 1993.

Luttwak, Edward. *Strategy: The Logic of War and Peace.* Cambridge, Mass.: Harvard University Press, 1987.

———. "A Post-Heroic Military Policy." *Foreign Affairs* 75, no. 4 (1996): 33–44.

Luvaas, Jay. *The Education of an Army: British Military Thought, 1815–1940.* London: Cassell, 1965.

Lyttelton, Adrian. *The Seizure of Power: Fascism in Italy, 1919–1929.* Weidenfeld and Nicolson, 1973.

Macfie, A. L. *The End of the Ottoman Empire, 1908–1923.* New York: Longman, 1998.

MacIsaac, David. "Voices from the Central Blue: The Air Power Theorists." In *Makers of Modern Strategy: From Machiavelli to the Nuclear Age,* edited by Peter Paret, 624–47. Princeton, N.J.: Princeton University Press, 1986.

MacIver, R. M. *The Modern State.* Oxford: Clarendon Press, 1926.

MacKenzie, John M. *Propaganda and Empire: The Manipulation of British Public Opinion, 1880–1960.* Manchester: Manchester University Press, 1984.

MacMillan, Margaret. *Paris 1919: Six Months That Changed the World.* New York: Random House, 2001.

Mahan, A. T. "The Place of Force in International Relations." *North American Review* 195 (January–June 1912): 28–39.

Maier, Charles. *Dissolution: The Crisis of Communism and the End of East Germany.* Princeton, N.J.: Princeton University Press, 1997.

Mandelbaum, Michael. *The Dawn of Peace in Europe.* New York: Twentieth Century Fund, 1996.

———. "Is Major War Obsolete?" *Survival* 40, no. 4 (1998–99): 20–38. Responses in *Survival* 41, no. 2 (1999): 139–52.

Markovits, Andrei. *Amerika, dich hasst sich's besser: Antiamerikanismus und Antisemitismus in Europa.* Hamburg: KVV konkret, 2004.

Marrin, Albert. *Sir Norman Angell.* Boston: Twayne, 1979.

Marrus, Michael. *The Unwanted: European Refugees in the Twentieth Century.* New York: Oxford University Press, 1985.

Martin, Michel. *Warrior to Managers: The French Military Establishment Since 1945.* Chapel Hill: University of North Carolina Press, 1981.

Maude, Frederic. *War and the World's Life.* London: Smith, Elder, 1907.

Maxwell, Kenneth. *The Making of Portuguese Democracy.* Cambridge: Cambridge University Press, 1995.

Mayer, Arno J. "The Domestic Origins of the First World War." In *The Responsibility of Power,* edited by F. Stern and L. Krieger. Garden City, N.Y.: Doubleday, 1967.

Mayer, J. P. *Max Weber and German Politics.* London: Faber, 1956.

Mazower, Mark. *Dark Continent: Europe's Twentieth Century.* New York: Knopf, 1998.

———. *The Balkans: A Short History.* New York: Modern Library, 2000.

McDonald, David. *United Government and Foreign Policy in Russia, 1900–1914.* Cambridge, Mass.: Harvard University Press, 1992.

Mearsheimer, John J. "Back to the Future: Instability in Europe After the Cold War." *International Security* 15, no. 1 (1990): 5–56.

Millon, Charles. "France and the Renewal of the Atlantic Alliance." *NATO Review,* no. 3 (May 1996): 13–16.

Milward, Alan. *The Reconstruction of Western Europe, 1945–1951.* Berkeley: University of California Press, 1984.

———. *The European Rescue of the Nation-State.* 2nd ed. London: Routledge, 2000.

Mombauer, Annika. *Helmuth von Moltke and the Origins of the First World War.* Cambridge: Cambridge University Press, 2001.

Moran, Lord. *The Anatomy of Courage.* London: Constable, 1946.

Moravcsik, Andrew. *The Choice for Europe: Social Purpose and State Power from Messina to Maastricht.* Ithaca, N.Y.: Cornell University Press, 1998.

Morgan, Glyn. *The Idea of a European Superstate: Public Justification and European Integration.* Princeton, N.J.: Princeton University Press, 2005.

Morrow, John. *German Air Power in World War I.* Lincoln: University of Nebraska Press, 1982.

Mosca, Gaetano. *The Ruling Class (Elementi di scienza politica).* New York: McGraw-Hill, 1939. Originally published in 1896.

Moskos, Charles, J. A. Williams, and David Segal, eds. *The Postmodern Military: Armed Forces After the Cold War.* New York: Oxford University Press, 2000.

Mosse, George L. *Fallen Soldiers: Reshaping the Memory of the World Wars.* New York: Oxford University Press, 1990.

Mueller, John. *Retreat from Doomsday: The Obsolescence of Major Wars.* New York: Basic Books, 1989.

———. *The Remnants of War.* Ithaca, N.Y.: Cornell University Press, 2004.

Murray, Williamson. *The Change in the European Balance of Power, 1938–1939: The Path to Ruin.* Princeton, N.J.: Princeton University Press, 1984.

Naimark, Norman. *The Russians in Germany: A History of the Soviet Zone of Occupation, 1945–1949.* Cambridge, Mass.: Harvard University Press, 1995.

———. *Fires of Hatred: Ethnic Cleansing in Twentieth-Century Europe.* Cambridge, Mass.: Harvard University Press, 2001.

———. "Stalin and Europe in the Postwar Period, 1945–1953: Issues and Problems." *Journal of Modern European History* 2, no. 1 (2004): 28–56.

Nationalgalerie Berlin. Das XIX. Jahrhundert. Katalog der ausgestellten Werke. Berlin: SMPK, 2002.

NATO Web site: www.nato.int.

Noakes, J., and G. Pridham. *Nazism, 1933–1945,* vol. 3, *Foreign Policy, War, and Racial Extermination.* Exeter: University of Exeter Press, 1995.

Nye, R. *The Origins of Crowd Psychology: Gustave le Bon and the Crisis of Mass Democracy in the Third Republic.* London: Sage, 1975.

O'Day, Alan. *Dimensions of Irish Terrorism.* New York: G. K. Hall, 1994.

Odom, William E. *The Collapse of the Soviet Military.* New Haven: Yale University Press, 1998.

Olivier, Frits, and Ger Teilter. "Democracy and the Armed Forces: The Dutch Experiment." In *Armed Forces and the Welfare Societies: Challenges in the 1980s,* edited by G. Harries-Jenkins, 54–95. New York: St. Martin's Press, 1983.

O'Loughlin, John. "The Political Geography of Conflict: Civil Wars in the Hegemonic Shadow." In *Geography of War and Peace: From Death Camps to Diplomats*, edited by Colin Flint, 85–110. New York: Oxford University Press, 2005.

Oram, Gerard. *Military Executions During World War One*. New York: Palgrave, 2003.

Ousby, Ian. *The Road to Verdun: France, Nationalism, and the First World War*. Garden City, N.Y.: Doubleday, 2002.

Overmans, Rüdiger. *Deutsche militärische Verluste im zweiten Weltkrieg*. Munich: R. Oldenbourg, 1999.

Overy, Richard. *The Air War, 1939–1945*. London: Constable, 1980.

———. *Why the Allies Won*. New York: Norton, 1995.

———. *Russia's War*. New York: Penguin, 1997.

Packer, George. *The Assassin's Gate: America in Iraq*. New York: Farrar, Straus and Giroux, 2005.

Paret, Peter. *Makers of Modern Strategy: From Machiavelli to the Nuclear Age*. Princeton, N.J.: Princeton University Press, 1986.

Paris, Michael. "The First Air Wars: North Africa and the Balkans, 1911–1913." *Journal of Contemporary History* 26 (1991): 97–109.

Pedroncini, Guy, et al. *Histoire militaire de la France*, vol. 3, *De 1871 à 1940*. Paris: Presses Universitaires de France, 1992.

Porter, Bernard. *The Absent-Minded Imperialists: What the British Really Thought About the Empire*. New York: Oxford University Press, 2005.

Porter, Bruce. *War and the Rise of the State: The Military Foundations of Modern Politics*. New York: Free Press, 1994.

Powaski, Ronald. *The Cold War: The United States and the Soviet Union, 1917–1998*. New York: Oxford University Press, 1998.

Preston, Paul. *The Triumph of Democracy in Spain*. London: Methuen, 1986.

The Proceedings of the Hague Peace Conferences. 5 vols. New York: Oxford University Press, 1920–21.

Ralston, David. *The Army of the Republic: The Place of the Military in the Political Evolution of France, 1871–1914*. Cambridge, Mass.: MIT Press, 1967.

Reynolds, David. "Great Britain and the Security 'Lessons' of the Second World War." In *The Quest for Stability: Problems of West European Security, 1918–1957*, edited by R. Ahmann, A. M. Birke, and M. Howard, 299–325. Oxford: Oxford University Press, 1993.

———. "1940: Fulcrum of the Twentieth Century?" *International Affairs* 66, no. 2 (1990): 325–50.

Rich, Norman. *Friedrich von Holstein: Politics and Diplomacy in the Era of Bismarck and Wilhelm II*. 2 vols. Cambridge: Cambridge University Press, 1965.

Rifkin, Jeremy. *The European Dream: How Europe's Vision of the Future Is Quietly Eclipsing the American Dream*. New York: Jeremy P. Tarcher/Penguin, 2004.

Romano, Sergio. *La quarta sponda: la guerra di Libia, 1911–1912.* Milan: Bompiani, 1977.

Ropp, Theodore. *War in the Modern World.* New York: Collier, 1962.

Rosanvallon, Pierre. *The New Social Question: Rethinking the Welfare State.* Princeton, N.J.: Princeton University Press, 2000.

Rossino, Alexander. *Hitler Strikes Poland: Blitzkrieg, Ideology, and Atrocity.* Lawrence: University Press of Kansas, 2003.

Rotberg, Robert, ed. *When States Fail: Causes and Consequences.* Princeton, N.J.: Princeton University Press, 2004.

Rothenberg, Gunther E. *The Army of Francis Joseph.* West Lafayette, Ind.: Purdue University Press, 1976.

Ruane, Kevin. *The Rise and Fall of the European Defense Community: Anglo-American Relations and the Crisis of European Defense, 1950–55.* New York: St. Martin's Press, 2000.

Sangiovanni, Mette. "Why a Common Security and Foreign Policy Is Bad for Europe." *Survival* 45, no. 4 (2003): 193–206.

Sayer, Derek. "British Reaction to the Amritsar Massacre." *Past and Present,* no. 131 (1991): 130–64.

Schaffer, Ronald. *Wings of Judgment: American Bombing in World War II.* New York: Oxford University Press, 1985.

Schell, Jonathan. *The Unconquerable World: Power, Nonviolence, and the Will of the People.* New York: Metropolitan, 2003.

Schivelbusch, Wolfgang. *The Culture of Defeat: On National Trauma, Mourning, and Recovery.* New York: Metropolitan, 2003.

Schmitt, Carl. *Der Begriff des Politischen.* Berlin: Duncker and Humbolt, 1963. Written in 1932.

Schoenbaum, David. *Zabern 1913: Consensus Politics in Imperial Germany.* London: George Allen and Unwin, 1982.

Schuker, Stephen. *The End of French Predominance in Europe: The Financial Crisis of 1924 and the Adoption of the Dawes Plan.* Chapel Hill: University of North Carolina Press, 1976.

Schuman, Robert. "France and Europe." *Foreign Affairs* 31, no. 3 (April 1953): 349–60.

Scott, L. V. *Conscription and the Attlee Governments: The Politics and Policy of National Service, 1945–1951.* New York: Oxford University Press, 1993.

Semmel, B. *Liberalism and Naval Strategy: Ideology, Interest, and Sea Power During the Pax Britannica.* Boston: Allen and Unwin, 1986.

Shaw, Martin. *Post-Military Society: Militarism, Demilitarization, and War at the End of the Twentieth Century.* Philadelphia: Temple University Press, 1996.

Shephard, Ben. *A War of Nerves: Soldiers and Psychiatrists in the 20th Century.* Cambridge, Mass.: Harvard University Press, 2001.

Sherman, David. *The Construction of Memory in Interwar France.* Chicago: University of Chicago Press, 1999.

Shils, Edward, and Morris Janowitz. "Cohesion and Disintegration in the Wehr-
     macht in World War II." *Public Opinion Quarterly* 12, no. 2 (1948): 280–315.
Silber, Laura, and Allan Little. *Yugoslavia: Death of a Nation.* New York: Penguin,
     1997.
Silberner, Edmund. *The Problem of War in 19th-Century Economic Thought.* Prince-
     ton, N.J.: Princeton University Press, 1946.
Simms, Brendan. *Unfinest Hour: Britain and the Destruction of Bosnia.* London:
     Allen Lane, 2001.
Sloan, Elinor. *The Revolution in Military Affairs: Implications for Canada and* NATO.
     Montreal: Queen's University Press, 2002.
Smith, Leonard, and Stéphane Audoin-Rouzeau, eds. *France and the Great War.*
     Cambridge: Cambridge University Press, 1993.
Sontag, Raymond J. *A Broken World, 1919–1939.* New York: Harper & Row, 1971.
Soutou, Georges-Henri. *La guerre de cinquante ans: les relations est-ouest, 1943–*
     1990. Paris: Fayard, 2001.
Spencer, Herbert. *The Man Versus the State, with Six Essays on Government, Society,*
     and Freedom. Indianapolis: Liberty, 1981.
Spiers, Edward. *Haldane: An Army Reformer.* Edinburgh: Edinburgh University
     Press, 1980.
Steiner, Zara. *The Lights That Failed: European International History, 1919–1933.*
     Oxford: Oxford University Press, 2005.
Stevenson, David. *Armaments and the Coming of War, 1904–1914.* Oxford: Claren-
     don Press, 1996.
———. *Cataclysm: The First World War as Political Tragedy.* New York: Basic
     Books, 2004.
Stockholm International Peace Research Institute. *World Armaments and Disar-
     mament Yearbook.* Stockholm: Almquist and Wiksell, 1976–1985.
Stone, Norman. "Army and Society in the Habsburg Monarchy, 1900–1914." *Past
     and Present,* no. 33 (1966): 95–111.
Strachan, Hew. *The First World War,* vol. 1, *To Arms.* New York: Oxford Univer-
     sity Press, 2001.
———. "War and Society in the 1920s and 1930s." In *The Shadows of Total War:
     Europe, East Asia, and the United States, 1919–1939,* edited by R. Chickering
     and S. Förster, 35–54. Cambridge: Cambridge University Press, 2003.
Stromberg, Roland. *Redemption by War: The Intellectuals and 1914.* Lawrence,
     Kan.: Regents, 1982.
Stueck, William. *The Korean War: An International History.* Princeton, N.J.:
     Princeton University Press, 1995.
Suri, Jeremy. *Power and Protest: Global Revolution and the Rise of Détente.* Cam-
     bridge, Mass.: Harvard University Press, 2003.
Survey of International Affairs. London: Oxford University Press, 1920–1963.
Suttner, Bertha von. *Die Haager Friedenskonferenz: Tagebuchblätter.* Düsseldorf:
     Zwiebelzwerg, 1985.

Taubman, William. *Khrushchev: The Man and His Era.* New York: Norton, 2003.

Tilly, Charles. *Coercion, Capital, and European States, A.D. 990–1900.* Cambridge, Mass.: Basil Blackwell, 1990.

———. *The Politics of Collective Violence.* New York: Cambridge University Press, 2003.

Titmuss, Richard. *Problems of Social Policy.* London: H. M. Stationery Office, 1950.

Todorov, Tzvetan. *Hope and Memory: Lessons from the Twentieth Century.* Princeton, N.J.: Princeton University Press, 2003.

Trachtenberg, Marc. *A Constructed Peace: The Making of the European Settlement, 1945–1963.* Princeton, N.J.: Princeton University Press, 1999.

———. "The Making of a Political System: The German Question in International Politics, 1945–1963." In *From War to Peace: Altered Strategic Landscapes in the Twentieth Century,* edited by Paul Kennedy and William I. Hitchcock, 103–19. New Haven: Yale University Press, 2000.

Travers, T.H.E. "Technology, Tactics, and Morale: Jean de Bloch, the Boer War, and British Military Theory, 1900–1914." *Journal of Modern History* 51 (1979): 264–86.

Treadgold, Donald. *Twentieth-Century Russia.* 2nd ed. Chicago: Rand McNally, 1964.

Treitschke, Heinrich von. *Politik.* 2 vols. 5th ed. Leipzig, 1922.

Ullman, Joan. *The Tragic Week: A Study of Anticlericalism in Spain, 1875–1912.* Cambridge, Mass.: Harvard University Press, 1968.

Van Creveld, Martin. *Fighting Power: German and U.S. Army Performance, 1939–1945.* London: Arms and Armour, 1983.

———. *Command in War.* Cambridge, Mass.: Harvard University Press, 1985.

———. *Technology and War: From 2000 B.C. to the Present.* New York: Free Press, 1989.

Vandervort, Bruce. *Wars of Imperial Conquest in Africa, 1830–1914.* Bloomington: Indiana University Press, 1998.

Verhey, Jeffrey. *The Spirit of 1914: Militarism, Myth, and Mobilization in Germany.* Cambridge: Cambridge University Press, 2000.

Vogel, Jakob. *Nationen im Gleichschritt: der Kult der "Nation in Waffen" in Deutschland und Frankreich, 1871–1914.* Göttingen: Vandenhoeck and Ruprecht, 1997.

Wæver, Ole. "Insecurity, Security, and Asecurity in the West European Non-War Community." In *Security Communities,* edited by Emmanuel Adler and Michael Barnett, 69–118. Cambridge: Cambridge University Press, 1998.

Wall, Irwin. *France, the United States, and the Algerian War.* Berkeley: University of California Press, 2001.

Waltz, Kenneth. "The Emerging Structure of International Politics." *International Security* 18, no. 2 (1993): 44–79.

Walzer, Michael. "On the Role of Symbolism in Political Thought." *Political Science Quarterly* 82 (June 1967): 191–204.

Watt, Donald Cameron. *How War Came: The Immediate Origins of the Second World War, 1938–1939*. New York: Pantheon, 1989.

Weber, Eugen. *Peasants into Frenchmen: The Modernization of Rural France, 1870–1914*. Stanford, Cal.: Stanford University Press, 1976.

———. *The Hollow Years: France in the 1930s*. New York: Norton, 1994.

Wehler, Hans-Ulrich. *Krisenherde des Kaiserreichs, 1871–1918*. 2nd ed. Göttingen: Vandenhoeck and Ruprecht, 1979.

Weinberg, Gerhard. *A World at Arms*. New York: Cambridge University Press, 1994.

Weiner, Amir. *Making Sense of War: The Second World War and the Fate of the Bolshevik Revolution*. Princeton, N.J.: Princeton University Press, 2001.

Weizsäcker, Richard von. *Deutsche Leidenschaft. Reden des Bundespräsidenten*. Stuttgart: Deutsche Verlags-Anstalt, 1994.

Wells, H. G. *The War of the Worlds*. London: W. Heinemann, 1898.

———. *First and Last Things: A Confession of Faith and Rule of Life*. Leipzig: Bernhard Tauschnitz, 1909.

Werner, Andrzej. "Bloch the Man: A Biographical Appreciation." In *The Future of War*, edited by G. Prins and H. Tromp. The Hague: Kluwer Law, 2000.

Whalen, Robert. *Bitter Wounds: German Victims of the Great War, 1914–1918*. Ithaca, N.Y.: Cornell University Press, 1984.

Wheeler-Bennett, John. *Munich: Prologue to Tragedy*. New York: Viking, 1964.

Wildman, Allan. *The End of the Russian Imperial Army*. 2 vols. Princeton, N.J.: Princeton University Press, 1980–1987.

Wilkinson, Spenser. *Britain at Bay*. London: Constable, 1910.

Winter, Denis. *Death's Men: Soldiers of the Great War*. London: Allen Lane, 1978.

Winter, Jay. *Sites of Memory, Sites of Mourning: The Great War in European Cultural History*. New York: Cambridge University Press, 1995.

Wolf, Charles, and Benjamin Zycher. *European Military Prospects, Economic Constraints, and the Rapid Reaction Force*. Santa Monica, Cal.: Rand Corp., 2001.

Woodward, E. L., and Rohan Butler, eds., *Documents on British Foreign Policy, 1919–1939*. London: H. M. Stationery Office, 1946– .

Wright, Gordon. *The Ordeal of Total War, 1939–1945*. New York: Harper & Row, 1968.

Yakovlev, Alexander N. *A Century of Violence in Soviet Russia*. New Haven: Yale University Press, 2002.

Zeldin, Theodore. *France, 1848–1945*. 2 vols. Oxford: Clarendon Press, 1973–1977.

Zielonka, Jan. *Europe as Empire: The Nature of the Enlarged European Union*. Oxford: Oxford University Press, 2006.

Zuber, Terence. "The Schlieffen Plan Reconsidered." *War in History* 6, no. 3 (1999): 262–305.

Index